Rational Expectations Econometrics

UNDERGROUND CLASSICS IN ECONOMICS

Progress in economics takes place in many different forums. Many of these forums are not "above ground," in the sense of published work freely available. They are in the form of unpublished notes, dissertations, government reports, and lectures. This series is dedicated to making the best of this underground literature more widely available to libraries, scholars, and their students. In so doing it will serve to fill a large gap in the contemporary economic literature.

Entries in this series can be on any topic of interest to economists, and they may occasionally be in rough or unfinished form. The only criteria are that the works have been influential, widely cited, of exceptional excellence, and, for whatever reason, never before published.

Other Titles in This Series

Behind the Diffusion Curve: Theoretical and Applied Contributions to the Microeconomics of Technology Adoption, Paul A. David

The Allocation of Scarce Resources: Experimental Economics and the Problem of Allocating Airport Slots, David M. Grether, R. Mark Isaac and Charles R. Plott

Lectures on Game Theory, Robert J. Aumann

Notes on the Theory of Choice, David M. Kreps

Functional Form and Utility: A Review of Consumer Demand Theory, Arthur S. Goldberger

Rational Expectations Econometrics

Lars Peter Hansen
and Thomas J. Sargent

WITH CONTRIBUTIONS BY
John Heaton, Albert Marcet
and William Roberds

Westview Press
BOULDER • SAN FRANCISCO • OXFORD

Underground Classics in Economics

This Westview softcover edition is printed on acid-free paper and bound in library-quality, coated covers that carry the highest rating of the National Association of State Textbook Administrators, in consultation with the Association of American Publishers and the Book Manufacturers' Institute.

Published in 1991 in the United States of America by Westview Press, Inc., 5500 Central Avenue, Boulder, Colorado 80301, and in the United Kingdom by Westview Press, 36 Lonsdale Road, Oxford OX2 7EW

Library of Congress Cataloging-in-Publication Data
Hansen, Lars Peter.
Rational expectations econometrics / by Lars Peter Hansen and Thomas J. Sargent.
p. cm.—(Underground classics in economics)
ISBN 0-8133-7800-1
1. Rational expectations (Economic theory). 2. Econometrics.
I. Sargent, Thomas J. II. Title. III. Series.
HB3730.H284 1991
330′.01′5195—dc20

89-16674
CIP

Printed and bound in the United States of America

The paper used in this publication meets the requirements of the American National Standard for Permanence of Paper for Printed Library Materials Z39.48-1984.

10 9 8 7 6 5 4 3 2 1

Contents

Acknowledgments

We are very grateful to Maria Bharwada for typesetting this manuscript in TeX. She did a wonderful job in converting our scribblings into the forms that appear in this book. Several of the chapters in this volume originally appeared as Working Papers or Staff Reports of the Research Department of the Federal Reserve Bank of Minneapolis. We would like to thank the Minneapolis Fed for supporting much of the research that led to these papers. Some of the more recent research was funded by grants from the National Science Foundation to Hansen and Sargent. We used earlier versions of Chapter 2 in classes and benefited from comments from students in those classes. Philip Braun and Narayana Kocherlakota provided some particularly helpful remarks. For computational assistance in Chapter 4 we thank Danny Quah and Ravi Jagannathan. A long time has passed since we could afford to hire either of them. We thank Andy Atkeson, John Heaton, Hsin Chang Lu, Robert E. Lucas, Jr., Preston Miller and Chi Wa Yuen for valuable comments on earlier drafts of Chapter 5. In addition, we are grateful to Hsin Chang Lu and Alex Taber for expert research assistance. Robin Lumsdaine read an earlier version of Chapter 7 and made some useful suggestions. Alex Taber was a big help in proof reading and criticizing the entire manuscript. Clark Burdick, Gregory Cumber, Kathy Glover and François Velde also assisted in preparing the manuscript and in helping write some TeX code.

Lars Hansen and Thomas Sargent

1

Introduction

by Lars Peter HANSEN and Thomas J. SARGENT

We doubt that the material in this book could be described as "classic", but so much of it has been 'underground' for so long that we seized the opportunity to publish it in the "Underground Classics" series.[1] The papers in this volume were written over a period of about twelve years, with versions of most of the papers being written between 1979 and 1982. These papers report some of our efforts to make the hypothesis of rational expectations econometrically operational. We delayed publishing these papers for one reason or another, mostly because we believed that some of the arguments could be improved or because we were too busy with other projects to put the finishing touches on these papers. We welcome the opportunity to publish these papers, rough as some of them are, in the "Underground Classics" series, whose editor Spencer Carr has told us that polish and finish would only detract from the underground flavor.

We are macroeconometricians, and have come at the hypothesis of rational expectations from the perspective of macroeconometrics. In the last two decades, the hypothesis of rational expectations has invaded economics from a variety of sources, including game theory and general equilibrium theory. But in macroeconomics, the first invaders were time series econometricians, who in the late 1960's were seeking methods of restricting the parameters of lag distributions in their econometric models.[2] In the late 1970's, the focus of attention changed from restricting distributed lags to restricting vector autoregressions. The papers in this volume are intended to carry on and contribute to this tradition.

For macroeconometricians, an exciting aspect of rational expectations models is the way their solution (or equilibrium) restricts an entire stochastic process of observables. Subsumed within the restrictions are explicit models of the various sources of 'disturbances' or errors in economic relations, aspects of econometric models that drive estimation and inference, but about which pre-rational expectations (mostly nonstochastic) macroeconomic theories had typically been silent. Rational expectations modelling promised to tighten the link between *theory* and *estimation*, because the objects produced by the theorizing are exactly the objects in terms of which econometrics is cast, e.g., covariance generating functions, Markov processes, and ergodic distributions.

Work on rational expectations econometrics has divided into two complementary but differing lines. The first line aims more or less completely to characterize the restrictions that a model imposes on a vector stochastic process of observables, and to use those restrictions to guide efficient estimation. This line is a direct descendant of the full system approach to estimating simultaneous equations models. It aims to estimate all of the *deep parameters* of a model simultaneously by exploiting the cross-equation restrictions that a rational expectations model imposes on those parameters. The benefits of this line of attack are described by Hansen and Sargent (1980a) and Sargent (1981), principal ones being the following three: (i) the ability to handle a range of assumptions about stochastic error processes and about decision variables that are present in the model but missing in the econometrician's data set (e.g., effort); (ii) the promise of estimating the full range of parameters required to study responses to various policy interventions; and (iii) the econometric advantages of increased efficiency that are associated with full system estimation methods.

The second line of work is the application of method of moments estimators to estimating the parameters that appear in the Euler equations associated with dynamic optimum problems. Hall (1978) and Hansen and Singleton (1982) recognized that if strong assumptions are made about those processes in an Euler equation that are unobservable to the econometrician (most often, that they are not present), then an Euler equation implies that a set of orthogonality conditions hold for the econometrician's data set. These orthogonality conditions can contain enough information to identify the parameters in the return function corresponding to that Euler equation. Thus, this approach holds the promise of estimating some parameters without the need to estimate (or indeed even to specify) a complete equilibrium model.

The second line of work has the advantage that it is easier to implement than the first. Its disadvantages are the restrictive assumptions on unobservables needed to validate the approach, and the fact that even when it is applicable, it does not attempt to estimate the full range of parameters that are typically required to analyze an interesting range of policy interventions. This second approach has caught on more than the first in applied work, undoubtedly mainly because of its ease.

Most of the papers in this book are contributions to the first line of work (though the third, fourth, and fifth chapters can be viewed as embracing a strategy that is somewhere in between the two lines). All of the papers provide ways of interpreting and restricting *vector autoregressions* in the light of some version of a rational expectations or equilibrium model. The economic theories used in this book are without exception linear rational expectations models. We chose this class purposefully because they match up so naturally with vector autoregressions. Prices and quantities in these models are determined by the interactions of agents who are optimally responding to disturbances informing them about their current and future prospects. Underlying several of the papers in this volume is a recurring theme about the link between the innovations in a vector autoregression and the disturbances to agents' information sets, a theme to which we now turn.

The Multitude of Moving Average Representations

Chapter 2 by Hansen and Sargent describes elements of the theory of linear least squares prediction. This chapter is just a set of lecture notes we have used to teach least squares theory to our students. Much of the discussion in the remainder of the book is cast in terms of the objects defined in this theory. Linear least squares prediction theory studies a vector stochastic process by decomposing it into two orthogonal pieces: a part that can be forecast as a linear function of its own past values, and a part that cannot be forecast. A key construction of the theory is Wold's decomposition theorem. One starts with an $(n \times n)$ positive semi-definite matrix sequence $\{C_x(\tau)\}$ and interprets $C_x(\tau)$ as the covariance $Ex_t x'_{t-\tau}$ for an $n \times 1$ vector stochastic process $\{x_t\}$. From the information in $\{C_x(\tau)\}$, one constructs a sequence of *projections* or *autoregressions* of x_t onto a linear space spanned by $(x_{t-1}, \ldots, x_{t-n})$. By studying the behavior of these projections as $n \to \infty$, one arrives at a decomposition that expresses x_t as the sum of the part that can be predicted linearly from past values and an *innovation* u_t that is orthogonal to the predictable part. By construction, the innovation process u_t is serially uncorrelated and lies in the space spanned by current and lagged x's (i.e., it is a forecast error). Under some additional assumptions, x_t can be expressed as a *moving average* of the innovation u_t, namely,

$$x_t = \sum_{j=0}^{\infty} D_j u_{t-j} . \tag{1}$$

This is known as the *Wold moving average representation* for x_t, and the $\{u_t\}$ process is said to be a *fundamental* white noise process for x_t. The term *fundamental* denotes that u_t is a white noise in terms of which x_t possesses a one-sided moving-average representation, and that u_t lies in the linear space spanned by current and lagged x_t's.

By virtue of the construction that leads to the Wold representation, the $\{u_t\}$ process is a white noise that corresponds to an innovation in an infinite order vector autoregression. Thus, the "innovation accountings" of Sims (1980) are statements about a moving-average representation (1).

Moving average representations are not unique for two distinct reasons. First, one can always multiply u_t in (1) by a nonsingular matrix U and obtain another Wold moving-average representation, namely,

$$x_t = \sum_{j=0}^{\infty} (D_j U^{-1}) (U u_{t-j}) , \tag{2}$$

in which $(U u_t)$ is the new innovation and $(D_j U^{-1})$ is the new impulse response function. This kind of nonuniqueness is the type confronted by Sims (1980) in his discussion of alternative orthogonalization schemes for the innovation process. It leaves the optimal predictions implied by all of the associated Wold moving average representations and the information in the vector $U u_t$ unaffected. Since U is nonsingular, the history of $U u_t$'s spans exactly the same linear space as does the history of u_t's (or the history of x_t's).

The second type of nonuniqueness of (1) does affect the information content of the residuals. There exists a family of other moving average representations

$$x_t = D^* (L) u_t^* , \tag{3}$$

where

$$D^* (L) = \sum_{j=0}^{\infty} D_j^* L^j , \quad \sum_{j=0}^{\infty} \text{ trace } D_j^* D_j^{*\prime} < +\infty ,$$

and where $D^* (L)$ satisfies

$$D(z) E u_t u_t' D(z^{-1})' = D^*(z) E u_t^* u_t^{*\prime} D^* (z^{-1})' , \quad |z| = 1 . \tag{4}$$

In (3), $\{u_t^*\}$ is an $(m \times 1)$ white noise, where $m \geq n$, and $D^*(L)$ is an $n \times m$ matrix polynomial in the lag operator L. Given $D(z)$, any $D^*(z)$ that satisfies (3) defines a moving average representation with associated vector white noise $\{u_t^*\}$. For *most* such representations, namely all non Wold representations, the history of u_t^* spans a *larger* space than the space generated by the corresponding history of x_t's. In general, (3) is a representation in which the history of x_t's fails to *reveal* the corresponding history of u_t^*'s.

Much of this book involves studying settings in which an economic model has an equilibrium that is most naturally represented in the form of a moving average representation of the type (3), but in which the internal structure of the model implies that this representation is not automatically a Wold representation. This poses a problem for interpreting the innovations u_t in a vector autoregression in terms of the innovations u_t^* that occur in the economic model. Some of the papers in this volume are concerned with characterizing the dimensions of this interpretation problem within particular concrete contexts. Other papers are concerned with providing methods for circumventing the problem conceptually and econometrically.

Exact Linear Rational Expectations Models

The paper "Exact Linear Rational Expectations Models" studies a class of models in which we immediately have to confront the multiplicity of moving average representations associated with a given stationary stochastic process. This paper studies a special class of linear rational expectations models that are constructed out of two sorts of relationships. Economic theory enters only in the first set of relationships, which consist of a set of linear relationships involving *only* current and past values of a subset of variables x_t observable by the econometrician, and expectations of future values of those observable variables. The qualifier "only" defines what we mean by an *exact* model – there are no disturbance terms in this relationship from econometrician's point of view. The fact that the econometrician has access to a smaller information set than do the agents is the only possible source of an econometric error term in this relationship. The second set of relationships are informational in nature, being the piece of a complete moving-average representation that governs the subset of observables x_t being forecast by the agents in the model. The strategy in this paper is to deduce the restrictions on the moving average of the *entire* vector x_t implied by the hypothesis of rational expectations. It turns out to be easy to write down these restrictions, and to describe strategies for estimating moving average representations subject to them.

This model formulation strategy naturally poses the following question: given that the model is correct, to which of multiplicity of moving-average representations do the rational expectations restrictions apply? In answering this question, we in effect characterize the extent of the problem of identifying moving-average representations consistent with an *exact* rational expectations model. This characterization has implications for likelihood-based procedures for estimation and inference because it indicates that the likelihood function will have multiple peaks corresponding to different moving-average representations. Despite this identification problem, the restrictions implied by these models are testable because the underidentified aspects pertain only to the flow of new information and not to the way the information restricts the covariances across variables.

Section 4 of "Exact Linear..." has some curiosity value. It introduces nonstationarities in the time series and then demonstrates the sense in which the variables are *cointegrated*, although we did not originally apply that term because we were unaware of the work of Granger and co-workers at the time that our paper was initially drafted.

The solution strategy employed in "Exact Linear..." has much in common with that used by Whiteman (1983). Whiteman extensively explores how to impose the rational expectations restrictions on the moving average representations for models that are *fully specified*, in the sense that as many relationships are specified as variables that are determined by the model. In contradistinction our exact linear rational expectations models are incomplete in that the laws of motion for the information variables and variables being forecast are not completely specified. Despite this difference in economic interpretation between our work and Whiteman's, Whiteman's work shows that many of the mathematical methods used in "Exact Linear..." can be useful in formulating complete models.

The class of models in "Exact Linear..." has a number of applications, but is quite special in that the theory must come in the form of an *exact* model, which means that the econometric model can contain no source of error *except* that the econometrician conditions on a smaller information set in forming forecasts than do economic agents.[3] The next chapter, "Two Difficulties in Interpreting Vector Autoregressions", discusses a more general setting in which there are additional sources of disturbances in econometric relations.

Two Difficulties in Interpreting Vector Autoregressions

One common case in which the version of (3) delivered by economic theory fails to match up with the Wold representation occurs when the number of shocks m impinging on agents' information exceeds the number n of processes observed by the econometrician. In this case, the u_t's are bound to summarize and confound the effects of the u_t's. However, the problem can emerge even when $m = n$. The paper "Two Difficulties in Interpreting Vector Autoregressions" studies the problem in two distinct contexts in which $n = m$. The first context is that of a dynamic equilibrium model in which two shocks are impinging on agents' information sets, namely, *supply* and *demand* or *endowment* and preference shocks. The equilibrium of the model is represented as a stochastic process for price and quantity that is a moving average of the shocks in agents' information sets (i.e., the u_t's). The paper studies how these shocks are related to the innovations in a vector autoregression for price and quantity. The paper describes how to express the u_t's as distributed lags of the u_t^*'s. When these distributed lags are not concentrated at zero lag, the $\{u_t\}$ process contains less information than does the $\{u_t^*\}$ process.[4]

The second context studied in the "Two Difficulties..." chapter is that of aggregation over time. The problem here is that the econometrician has data sampled at a coarser interval than that of economic agents. Taking this idea to the limit, suppose that economic agents receive and process information in continuous time, and that an economic theory is in the form of a continuous time version of (3). (Think of a limiting version of (3) approached by successively reformulating (3) at finer and finer sampling intervals). Under what circumstances will the impulse response functions from the vector autoregression associated with a discrete time, sampled version of the x_t process resemble in shape the impulse response function in continuous time? This is a version of the aggregation over time question of Sims (1971) and Geweke (1978), who studied how well discrete time distributed lags would approximate their underlying continuous time counterparts. Rather than studying how these distributed lags (projections of one variable on leads and lags of another) match up, we study how the moving-average representations match up. The paper describes some conditions on the continuous time stochastic process involving *smoothness* (i.e., mean square continuity and differentiability) under which a discrete time moving average must fail to match up well with the underlying continuous time one.

The "Two Difficulties..." chapter assumes that the underlying continuous time model takes the form of a rational spectral density, which automatically imposes continuity on the kernel defining the continuous time moving average representation. In Chapter 10, which was written by Albert Marcet, this continuity requirement is dropped. Marcet studies discontinuous continuous time moving average kernels, in particular, how they affect the closeness of the continuous and discrete time moving average representations. Marcet offers a useful characterization that allows him to create a number of interesting examples in which the continuous and discrete time moving-average representations diverge.

A failure of the moving averages associated with vector autoregressions to match up with those associated with an underlying economic theory suggests caution in interpreting innovation accountings based on estimates of a theoretical vector autoregressions. If a researcher is willing to impose sufficient economic theory on the process of econometric estimation, the difficulties described above can be overcome in the sense both that consistent and efficient parameter estimates can be obtained, and that estimates of the moving-average representation corresponding to the economic model can be computed. In discrete time, Hansen and Sargent (1980a, 1981a) described strategies for carrying out such estimation. Corresponding methods for continuous time were described by Hansen and Sargent in unpublished papers (1980b, 1981d). Versions of these continuous time methods are described in this volume in papers by Hansen, Heaton, and Sargent and by Hansen and Sargent.

Three Papers on Continuous Time Rational Expectations Models

Chapters 7, 8, and 9 study three problems that must be solved if a continuous time dynamic equilibrium model is to be estimated via maximum likelihood methods. Chapter 7 by Hansen, Heaton, and Sargent and chapter 8 by Hansen and Sargent are concerned with computing the solution of a continuous time model in a form designed for econometric tractability. These two papers exploit the *certainty equivalence* feature of linear dynamic models, namely, that their solution can typically be broken into two separate steps: "optimization" and "forecasting". In effect, Hansen, Heaton, and Sargent describe how the equilibrium of a class of deterministic continuous time linear equilibrium models can be computed by solving a linear quadratic optimal control problem. The paper describes how a fast algorithm (a matrix sign algorithm) can be put to work on this problem. The product of Hansen, Heaton, and Sargent's calculations is an equilibrium in *feedback part, feedforward part* form. This is the solution of a deterministic version of the model, in which future paths of forcing variables are known with certainty. To obtain the solution for stochastic versions of the model in which the forcing functions are stochastic processes, one simply substitutes linear least squares forecasts for future values in the feedforward part, leaving the feedback part unaltered. In chapter 8, Hansen and Sargent describe a set of convenient formulas for computing the feedforward parts in the stochastic case. These are the continuous time counterparts of formulas described by Hansen and Sargent (1980a, 1981b) for the discrete time case.

By combining the results in these two papers, one obtains a representation of the solution in the form of a vector linear stochastic differential equation. This representation provides a continuous time version of the solutions described by Hansen and Sargent (1980a, 1990). Given such a representation, standard methods can be used to compute the likelihood function of the continuous time model conditioned on discrete time data. These methods are described by Jones (1980), Ansley and Kohn (1983), Bergstrom (1983), Harvey and Stock (1985), and Hansen and Sargent (1980b). Christiano, Eichenbaum, and Marshall (1990) have applied such methods to study consumption smoothing models using U.S. time series data.

Chapter 9 by Hansen and Sargent treats a special case of the identification problem that must be solved in estimating a continuous time model from discrete time data. As described by Phillips (1973) and Hansen and Sargent (1981c), without some restrictions on the continuous time model, there is typically a multitude of continuous time models that is consistent with a given discrete time covariance generating function. This is a version of the classic *aliasing* phenomenon. Hansen and Sargent illustrate how the cross-equation restrictions imposed by rational expectations can serve to resolve this identification problem. While Hansen and Sargent obtain analytical results only for a special case, their results suggest numerical methods that can be used to check identification in more general models.

Testing Present Value Budget Balance

Chapter 5 by Hansen, Roberds, and Sargent describes a class of models in which the moving-average representation delivered by theory is not a Wold representation. Research on the subject of this paper was initiated in response to a question posed by Robert E. Lucas, Jr. at a conference in October 1985. Lucas asked what restrictions would imposed on a joint stochastic process describing net of interest government expenditures and taxes by the assumption of present value budget balance. Lucas conjectured that even with a constant real interest rate, the restriction would be a weak one because of the ability to postpone repayment now via an indefinite promise to run surpluses later. An incomplete analysis of the issue was made by Sargent (1987b), who showed that the hypothesis of present value budget balance imposes the restriction that the present value of the impulse response coefficients of the deficit to each innovation in agents' information set is zero. This is also one of the restrictions imposed by the consumption smoothing model used by Hall (1978) and the tax smoothing model used by Barro (1979). One of Hansen, Roberds, and Sargent's aims is to characterize exhaustively the restrictions implied by a class of models including Hall's and Barro's as special cases.

Let g_t denote government expenditures and let τ_t denote tax collections, both net of interest. Hansen, Roberds, and Sargent start by studying the case in which observations on g_t, τ_t, but *not* on the debt, are available to the econometrician. They assume a constant real interest rate. They begin with the observation that, in general, the restriction that the present value of the moving-average coefficients for the deficit $\tau_t - g_t$ be zero implies that that moving-average cannot be a Wold representation. The reason is that the restriction itself implies that the moving-average polynomial in the lag operator is not invertible. This has the implication that the restriction ought not to be tested by checking whether it holds for the impulse response function associated with a vector autoregression (i.e., a Wold representation).

Hansen, Roberds, and Sargent go on to show that the restriction itself is vacuous unless additional restrictions are imposed on the moving-average representation. In particular, they show that given a moving-average for $\{(g_t, \tau_t)\}$ that violates the restriction, one can always find another moving-average representation that satisfies the restriction. To build this alternative representation, one just needs sufficient flexibility in the parameterization of the alternative moving-average representation. This result confirms Lucas's initial skepticism about the restrictiveness of the budget balance restriction.

Hansen, Roberds, and Sargent then describe two contexts in which the restriction is testable. The first context, described in sections 3-5 of their paper, imposes additional theoretical structure in the form of a version of an optimal consumption smoothing model. The second assumes that additional data are available, namely, a time series on the stock of real interest bearing debt.

In the first setting, Hansen, Roberds and Sargent study whether the hypothesis of present value budget balance adds any testable implications to a martingale model for the marginal utility of consumption. They show that it does add a testable restriction, and that a form of this additional restriction continues to hold in the case in which preferences are nonseparable in consumption. Hansen, Roberds, and Sargent show that the source of this testable restriction is the ability that the martingale model has to identify one component of agents' information set, namely, the innovation to *consumption outlays*, which the theory states is a linear combination of the innovations to agents' information sets.

In sections 4 and 5 of their paper, Hansen, Roberds and Sargent describe and implement a strategy for testing this implication of present value budget balance in the context of a martingale model with nonseparable preferences. Hansen, Roberds, and Sargent describe a *semi-parametric* testing strategy, in that they do not directly specify and estimate the various technology and preference parameters in their underlying model (as would be done, for example, if one were pursuing the estimation strategy for those models described in Hansen and Sargent 1990). Instead, they parameterize some particular lag distributions that are mongrel parameterizations of the structural parameters. This strategy is motivated by a desire to focus on the present value budget restriction while remaining noncommittal about details of the nonseparabilities and the number of goods in the model. Hansen, Roberds, and Sargent apply this test to U.S. data on consumption and labor income, and find little evidence against the present value budget restriction. It might be interesting to repeat this test for a generalized *tax smoothing* model using U.S. data.

However, Section 6 raises a cautionary note concerning the interpretation of these tests. Section 6 establishes the existence of a sequence of false models that satisfy the restrictions but that approximate data that violate the restriction arbitrarily well. This result has to dampen somewhat one's enthusiasm for our semi-parametric strategy, since it suggests that if we get nonparametric enough, the present value budget restriction becomes virtually vacuous even with the martingale model also imposed.[5]

Section 7 of the Hansen, Roberds, Sargent paper changes the setup in two ways, first to permit time varying interest rates and second to add observations on debt to the expenditure and revenue series. The paper shows that the present value budget balance restriction leads to an exact linear rational expectations model. In a separate paper by Roberds, such a model is estimated and the restriction is tested for U.S. data on government expenditures and taxes. Roberds finds evidence against the restriction for these data.

Notes

1. Chapter 2 was written in 1981–82. Chapter 3 was written in 1980–81 (Hansen and Sargent 1981e), and revised in 1990. Chapter 4 was written in 1982, with minor revisions being made in 1984 (Hansen and Sargent 1984) and 1989. Most of chapter 5 was written in 1987, with the empirical work being completed in 1990. Chapter 6 was written in 1988. Chapters 7 and 8 were written in 1988–90. They amount to revisions and extensions of ideas that initially appeared in working papers by Hansen and Sargent that appeared in 1980 and 1981 (Hansen and Sargent (1980b, 1981d). Chapter 9 was completed in 1980–81 (Hansen and Sargent 1981c). Chapter 10 was written as part of Albert Marcet's Ph.D. dissertation, which was completed in 1987.

2. See Muth (1960, 1961), Nerlove (1967), Griliches (1967), Sims (1974), Lucas (1972), and Sargent (1971).

3. This way of introducing errors in econometric models was used to great advantage by Shiller (1972).

4. The paper also describes how its partial equilibrium model is to be interpreted as a special case of the class of general equilibrium models studied by Hansen and Sargent (1990).

5. See Sargent (1987b, chapter XIII) for a discussion of the relationship between Hall's (1978) model and Barro's (1979) tax smoothing model. Evidently, there are tax smoothing models that are similarly related to the general class of consumption smoothing models described by Hansen, Roberds, and Sargent or by Hansen and Sargent (1990).

2

Lecture Notes on Least Squares Prediction Theory

by Lars Peter HANSEN and Thomas J. SARGENT

1. Introduction

In these notes we establish some basic results for least squares prediction theory. These results are useful in a variety of contexts. For instance, they are valuable for solving linear rational expectations models, representing covariance stationary time series processes, and obtaining martingale difference decompositions of strictly stationary processes.

The basic mathematical construct used in these notes is an inner product defined between two random variables. This inner product is calculated by taking the expectation of the product of the two random variables. Many of the results obtained using this particular inner product are analogous to results obtained using the standard inner product on multi-dimensional Euclidean spaces. Hence intuition obtained for Euclidean spaces can be quite valuable in this context as well.

The formal mathematical machinery that is exploited in these notes is the Hilbert space theory. There is a variety of references on Hilbert spaces that should provide good complementary reading, e.g. Halmos (1957) and Luenberger (1969).

2. Prediction Problem

In this section we specify formally the problem of forecasting a random variable y given a collection of random variables H. This problem is sufficiently general to include conditional expectations and best linear predictors as special cases. We also consider a second problem that is closely related to the prediction problem. This second problem is termed the orthogonality problem and can be interpreted as providing a set of necessary and sufficient first-order conditions for the prediction problem. In particular, we will show that these two problems have the same solution when some auxiliary assumptions are imposed on the set H.

A formal statement of the least squares prediction problem is:

PROBLEM 2.1 (Least Squares Forecasting Problem): Find the random variable h_o in H for which $E[(y - h_o)^2] \leq E[(y - h)^2]$ for all h in H.

There is a second problem that often has the same solution as Problem 2.1. This second problem uses the following definition in its statement.

Definition 2.1: The random variable y is orthogonal to the set H if $E(hy) = 0$ for all h in H.

The second problem is:

PROBLEM 2.2 (Orthogonality Problem): Find the random variable h_o in H such that $y - h_o$ is orthogonal to H.

To ensure that the statements of these two problems are well-defined, we make the assumption that the random variable being forecast and the random variables that are used in forecasting have finite second moments. With this in mind, we let L^2 denote the set of all random variables defined on the underlying probability space that have finite second moments. We assume that both y and the elements in H are in L^2. There are two important inequalities that apply to random variables in L^2.

Lemma 2.1: (Cauchy-Schwarz Inequality): For any y_1 and y_2 in L^2, $|E(y_1 y_2)| \leq E(y_1^2)^{\frac{1}{2}} E(y_2^2)^{\frac{1}{2}}$.

Proof: If either y_1 or y_2 is zero, then the inequality is satisfied trivially. Suppose that y_1 and y_2 are both different from zero. Then $E(y_1^2)$ and $E(y_2^2)$ are both positive. Notice that

$$
\begin{aligned}
0 &\leq \left[\frac{|y_1|}{(Ey_1^2)^{\frac{1}{2}}} - \frac{|y_2|}{(Ey_2^2)^{\frac{1}{2}}}\right]^2 \\
&= \left[\frac{y_1^2}{E(y_1^2)} - \frac{2|y_1 y_2|}{E(y_1^2)^{\frac{1}{2}} E(y_2^2)^{\frac{1}{2}}} + \frac{y_2^2}{E(y_2^2)}\right] .
\end{aligned} \tag{2.1}
$$

Consequently,

$$
\frac{|y_1 y_2|}{E(y_1^2)^{\frac{1}{2}} E(y_2^2)^{\frac{1}{2}}} \leq \frac{1}{2}\left[\frac{y_1^2}{E(y_1^2)} + \frac{y_2^2}{E(y_2^2)}\right] . \tag{2.2}
$$

Taking expectations of both sides of (2.2) gives

$$\frac{E|y_1 y_2|}{E(y_1^2)^{\frac{1}{2}} E(y_2^2)^{\frac{1}{2}}} \leq 1 . \tag{2.3}$$

Multiplying by $E(y_1^2)^{\frac{1}{2}}$ and $E(y_2^2)^{\frac{1}{2}}$ gives

$$E(|y_1 y_2|) \leq E(y_1^2)^{\frac{1}{2}} E(y_2^2)^{\frac{1}{2}} . \tag{2.4}$$

The Cauchy-Schwarz Inequality then follows from the fact that

$$|E(y_1 y_2)| \leq E(|y_1 y_2|) . \blacksquare \tag{2.5}$$

Lemma 2.2: (Triangle Inequality): For any y_1 and y_2 in L^2, $E[(y_1 + y_2)^2]^{\frac{1}{2}} \leq E(y_1^2)^{\frac{1}{2}} + E(y_2^2)^{\frac{1}{2}}$.

Proof: Notice that

$$\begin{aligned} (y_1 + y_2)^2 &= y_1^2 + 2 y_1 y_2 + y_2^2 \\ &\leq y_1^2 + 2|y_1 y_2| + y_2^2 . \end{aligned} \tag{2.6}$$

Taking expectations and using inequality (2.4) gives

$$\begin{aligned} E[(y_1 + y_2)^2] &\leq E(y_1^2) + 2E(y_1^2)^{\frac{1}{2}} E(y_2^2)^{\frac{1}{2}} + E(y_2^2) \\ &= [E(y_1^2)^{\frac{1}{2}} + E(y_2^2)^{\frac{1}{2}}]^2 . \end{aligned} \tag{2.7}$$

Finally, taking square roots of each side of (2.7) gives the Triangle Inequality. $\blacksquare$

Among other things, the Cauchy-Schwarz and Triangle Inequalities imply that the statements of Problems 2.1 and 2.2 are well-defined as long as y and the elements of H are in L^2. In addition, the Triangle Inequality implies that L^2 is a linear space in the sense that linear combinations of elements in L^2 are in L^2. When h_o solves Problem 2.2 for a subset of G of L^2, then h_o also solves Problem 2.2 for the set

$$\begin{aligned} H = \Big\{ h : h = c_1 g_1 + c_2 g_2 + \cdots + c_n g_n \text{ for some integer } n, \\ \text{some real numbers } c_1, c_2, \ldots, c_n, \text{ and some elements} \\ g_1, g_2, \ldots, g_n \text{ in } G \Big\} . \end{aligned} \tag{2.8}$$

Thus H is constructed to be a linear subspace of L^2.

Definition 2.2: The set H is a linear subspace of L^2 if H is a subset of L^2 and if for any h_1 and h_2 in H and any real numbers c_1 and c_2, the random variable $c_1h_1 + c_2h_2$ is in H.

Consequently, when h_o solves Problem 2.2 on an arbitrary subset of L^2, this subset can always be expanded to be a linear subspace of L^2 via (2.8). The random variable h_o will continue to solve Problem 2.2 on the expanded set. In what follows we will focus on linear subspaces of H.

Our next two lemmas show that Problems 2.1 and 2.2 have the same solution when H is a linear subspace of L^2.

Lemma 2.3: Suppose that y is in L^2 and H is a linear subspace of L^2. If h_o solves problem 2.2, then h_o is the unique solution to Problems 2.1 and 2.2.[1]

Proof: Let h_o be the solution to Problem 2.2. For any h in H,

$$
\begin{aligned}
E[(y-h)^2] = E[(y-h_o+h_o-h)^2] &= E[(y-h_o)^2]+ \\
2E[(y-h_o)(h_o-h)] &+ E[(h_o-h)^2] \\
&= E[(y-h_o)^2] + E[(h_o-h)^2] .
\end{aligned}
\tag{2.9}
$$

Hence,

$$E[(y-h_o)^2] \leq E[(y-h)^2] \tag{2.10}$$

for any h in H implying that h_o solves Problem 2.1. Furthermore, if $E[(h-h_o)^2]$ is greater than zero, then

$$E[(y-h_o)^2] < E[(y-h)^2] . \tag{2.11}$$

Since $E[(h_o-h)^2] = 0$ only if $h = h_o$, h_o is the unique solution to Problem 2.1. It follows that h_o is also the unique solution to Problem 2.2 since we have shown that any solution to Problem 2.2 is a solution to Problem 2.1. ∎

Lemma 2.4: Suppose y is in L^2 and H is a linear subspace of L^2. If h_o solves Problem 2.1, then h_o also solves Problem 2.2.

Proof: Let h_o be the solution to Problem 2.1 and let h be any element in H. If h is zero, it follows immediately that $E[(y-h_o)h]$ is zero. If h is different from zero, let $\delta = E[(y-h_o)h]/E(h^2)$. Notice that

$$
\begin{aligned}
E[(y-h_o-\delta h)^2] &= E[(y-h_o)^2] + \delta^2 E[h^2] - 2\delta E[(y-h_o)h] \\
&= E[(y-h_o)^2] - \delta^2 E[h^2] .
\end{aligned}
\tag{2.12}
$$

Since h_o solves Problem 2.1 and $h_o + \delta h$ is in H, δ must be zero. However, δ is zero only when $E[(y - h_o)h]$ is zero. Since our choice of h in H was arbitrary, $y - h_o$ is orthogonal to H. ∎

Lemmas 2.3 and 2.4 showed that Problems 2.1 and 2.2 have the same solution as long as H is a linear subspace of L^2. However, we have not shown that a solution exists for either problem. An additional restriction on the set H is sufficient to guarantee that a solution exists. This restriction uses the following definitions:

Definition 2.3: A sequence $\{y_n : n \geq 1\}$ in L^2 is Cauchy if $\lim_{n\to\infty} \sup_{m\geq n} E[(y_n - y_m)^2] = 0$.

Definition 2.4: A sequence $\{y_n : n \geq 1\}$ in L^2 is convergent if there exists a random variable y_o in L^2 for which $\lim_{n\to\infty} E[(y_n - y_o)^2] = 0$.[2]

The linear space L^2 is complete in the sense that any Cauchy sequence in L^2 is convergent (see the Appendix). If the analogous property holds for a subset of L^2, we say that the subset is closed.

Definition 2.5: A subset H of L^2 is closed if any Cauchy sequence in H converges to an element in H.

If a given set H is not closed, we can form its closure by adding all limit points of Cauchy sequences in H to H. The resulting augmented set will be closed.

Lemma 2.5: Suppose H is a closed linear subspace of L^2. Then, for any y in L^2, Problem 2.1 has a solution.

Proof: Let δ be the nonnegative real number that satisfies

$$\delta^2 = \inf_{h\in H} E[(y - h)^2] , \tag{2.15}$$

and let $\{h_n : n = 1, 2, \ldots\}$ be a sequence of random variables in H for which

$$\lim_{n\to\infty} E[(y - h_n)^2] = \delta^2 . \tag{2.16}$$

First, we show that when H is a linear space, $\{h_n : n \geq 1\}$ is a Cauchy sequence. Notice that

$$\begin{aligned} &E[(h_m - h_n)^2] + 4E\{[(1/2)(h_m + h_n) - y]^2\} \\ &\qquad = 2E[(h_m - y)^2] + 2E[(h_n - y)^2] . \end{aligned} \tag{2.17}$$

Therefore,

$$E[(h_m - h_n)^2] \leq 2E[(h_m - y)^2] + 2E[(h_n - y)^2] - 4\delta^2 \tag{2.18}$$

since $(1/2)(h_m + h_n)$ is in H and $\delta^2 < E[(y-h)^2]$ for any h in H. Taking limits of both sides of (2.18) gives

$$\begin{aligned} 0 \leq \lim_{m\to\infty} \sup_{m\geq n} E[(h_m - h_n)^2] &\leq 2 \lim_{m\to\infty} E[(h_m - y)^2] + \\ &\quad 2 \lim_{n\to\infty} E[(h_n - y)^2] - 4\delta^2 \\ &= 2\delta^2 + 2\delta^2 - 4\delta^2 \\ &= 0\,. \end{aligned} \tag{2.19}$$

Hence, $\{h_n : n \geq 1\}$ is Cauchy as long as H is a linear space. If in addition H is closed, this sequence converges to an element, h_o, in H. We now show that h_o is the solution to the least squares forecasting problem. By the Triangle Inequality (Lemma 2.2),

$$E[(y - h_o)^2]^{\frac{1}{2}} \leq E[(y - h_n)^2]^{\frac{1}{2}} + E[(h_n - h_o)^2]^{\frac{1}{2}}\,. \tag{2.20}$$

Taking limits of both sides of (2.20) gives

$$\left\{E[(y - h_o)^2]\right\}^{\frac{1}{2}} \leq \delta \tag{2.21}$$

which proves that h_o solves Problem 2.1. ∎

We have shown that if H is a closed linear space then there exists a unique solution to the least squares forecasting problem (Problem 2.1). In these circumstances we define the projection operator as follows.

Definition 2.6: For any H that is a closed linear subspace of L^2, the least squares projection operator, $P(y|H)$ is the solution to Problem 2.1 of forecasting y given H.

There are two trivial examples of projection operators. For the first example let H be L^2. The projection operator $P(\cdot|L^2)$ is just the identity operator since $P(y|L^2) = y$. For the second example let H be the set Z that contains only the zero random variable. The projection operator $P(\cdot|Z)$ maps all random variables into the zero random variable since the zero random variable is the only random variable in Z. In sections five and six we study two nontrivial projection operators that are used in studying time series prediction problems.

One can be interpreted as a best linear predictor and the other as a conditional expectation.

Our final result in this section establishes that $P(\cdot|H)$ satisfies a linearity property.

Lemma 2.6: Suppose that H is a closed linear subspace of L^2. Then for any y_1 and y_2 in L^2 and any real numbers c_1 and c_2

$$P(c_1y_1 + c_2y_2|H) = c_1P(y_1|H) + c_2P(y_2|H)\ . \tag{2.22}$$

Proof: To prove this lemma, we will show that the right-hand side of (2.22) solves the orthogonality problem (Problem 2.2). Notice that $y_1 - P(y_1|H)$ and $y_2 - P(y_2|H)$ are orthogonal to H. Consequently,

$$c_1y_1 + c_2y_2 - c_1P(y_1|H) - c_2P(y_2|H) \tag{2.23}$$

is orthogonal to H. Furthermore, $c_1P(y_1|H) + c_2P(y_2|H)$ is in H. Therefore, $c_1P(y_1|H) + c_2P(y_2|H)$ solves the orthogonality problem. ∎

3. Useful Results Involving Projections

In this section we present some results that are useful in calculating projections. In preparation for these results, we introduce some new concepts.

Definition 3.1: The sets G and U are orthogonal if $E(gu) = 0$ for all g in G and all u in U.

Definition 3.2: The sum of the sets G and U is $G + U = \{g + u : g$ is in G and u is in $U\}$.

The sum of two closed linear subspaces of L^2 that are orthogonal is also a closed linear subspace of L^2.

Lemma 3.1: Suppose that G and U are orthogonal closed linear subspaces of L^2. Then $G + U$ is a closed linear subspace of L^2.

Proof: It is straightforward to show that if G and U are linear spaces, then their sum is also a linear space. We leave this as an exercise for the reader. To show that $G + U$ is closed, let $\{(g_n + u_n) : n \geq 1\}$ be a Cauchy sequence in $G + U$ where $\{g_n : n \geq 1\}$ is a sequence in G and $\{u_n : n \geq 1\}$ is a sequence in U. We will show that these latter two sequences are Cauchy. Since $g_n - g_m$ is in G, $u_n - u_m$ is in U, and G and U are orthogonal,

$$E[(g_n + u_n - g_m - u_m)^2] = E[(g_n - g_m)^2] + E[(u_n - u_m)^2]\ . \tag{3.1}$$

Therefore, the following two inequalities hold:

$$E[(g_n + u_n - g_m - u_m)^2] \geq E[(u_n - u_m)^2] , \tag{3.2}$$

and

$$E[(g_n + u_n - g_m - u_m)^2] \geq E[(g_n - g_m)^2] . \tag{3.3}$$

Since $\{g_n + u_n : n \geq 1\}$ is Cauchy, it follows that $\{g_n : n \geq 1\}$ and $\{u_n : n \geq 1\}$ are Cauchy and hence convergent. Recall that both G and U are assumed to be closed so that $\{g_n : n \geq 1\}$ converges to an element g_o in G and $\{u_n : n \geq 1\}$ converges to an element u_o in U. Since

$$E[(g_n + u_n - g_o - u_o)^2] = E[(u_n - u_o)^2] + E[(g_n - g_o)^2] , \tag{3.4}$$

$\{g_n + u_n : n \geq 1\}$ converges to $g_o + u_o$ in $G + U$. ∎

Among other things Lemma 3.1 guarantees that the projection operator is well-defined on $G + U$ when G and U are orthogonal closed linear spaces. In this case the projection onto $G + U$ is the sum of the projections onto G and U.

Lemma 3.2: Suppose that G and U are orthogonal closed linear subspaces of L^2. Then $P(\cdot|G + U) = P(\cdot|G) + P(\cdot|U)$.

Proof: Let y be any element in L^2. To prove this lemma, we will show that $y - P(y|G) - P(y|U)$ is orthogonal to $G + U$. Since $P(y|G)$ solves the orthogonality problem (Problem 2.2) for the subspace G, $y - P(y|G)$ is orthogonal to G. Since U is assumed to be orthogonal to G, $P(y|U)$ is orthogonal to G. Hence, $y - P(y|G) - P(y|U)$ is orthogonal to G. By reversing the roles of G and U, it follows that $y - P(y|U) - P(y|G)$ is also orthogonal to U. Therefore, $y - P(y|G) - P(y|U)$ is orthogonal to $U + G$. ∎

Lemma 3.2 gives a convenient way for revising forecasts when new information is received. For instance, suppose that G contains information available in the past and U contains new information (orthogonal to G) that becomes available today. Then Lemma 3.2 shows how to update forecasts after the arrival of the new information. This raises the following question. When can new information be viewed as a closed linear space that is orthogonal to the set of information available previously? The following result will be used in answering this question.

Lemma 3.3: Suppose the sequences $\{y_n : n \geq 1\}$ and $\{y_n^* : n \geq 1\}$ in L^2 converge to y_o and y_o^*, respectively. Then the sequence of real numbers $\{E(y_n y_n^*) : n \geq 1\}$ converges to $E(y_o y_o^*)$.

Proof: Notice that

$$\begin{aligned} E(y_n y_n^*) - E(y_o y_o^*) &= E[y_o(y_n^* - y_o^*)] \\ &+ E[(y_n - y_o)y_o^*] \\ &+ E[(y_n - y_o)(y_n^* - y_o^*)] \, . \end{aligned} \tag{3.5}$$

From the Cauchy-Schwarz Inequality and the Triangle Inequality for real numbers,

$$\begin{aligned} \left| E(y_n y_n^*) - E(y_o y_o^*) \right| &\leq E(y_o^2)^{\frac{1}{2}} E[(y_n^* - y_o^*)^2]^{\frac{1}{2}} \\ &+ E[(y_n - y_o)^2]^{\frac{1}{2}} E(y_o^{*2})^{\frac{1}{2}} \\ &+ E[(y_n - y_o)^2]^{\frac{1}{2}} E[(y_n^* - y_o^*)^2]^{\frac{1}{2}} \, . \end{aligned} \tag{3.6}$$

The conclusion follows from noting that the right-hand side of (3.6) converges to zero since $\{y_n - y_o : n \geq 1\}$ and $\{y_n^* - y_o^* : n \geq 1\}$ converge to zero. ▮

Our next lemma gives a construction for the set U of new information.

Lemma 3.4: Suppose that G and H are closed linear subspaces of L^2 and G is a subset of H. Let

$$U = \{u : u = h - P(h|G) \text{ for some } h \text{ in } H\} \, . \tag{3.7}$$

Then U is a closed linear subspace of L^2 and $H = U + G$.

Proof: Since G is a subset of H, $h - P(h|G)$ is in H for any h in H. Therefore, U is a subset of H that is orthogonal to G. Hence, $U + G$ is a subset of H. However, any h in H can be represented as $h = P(h|G) + [h - P(h|G)]$ which shows that h is in $U + G$. Consequently, $H = G + U$ where G and U are orthogonal. Next we show that U is a closed linear subspace of L^2. Let u be any element in H that is orthogonal to G. Then $P(u|G) = 0$ so that u is in U. Since any element of U is orthogonal to G,

$$U = \{u \text{ in } H : u \text{ is orthogonal to } G\} \, . \tag{3.8}$$

Recall that H is a linear subspace of L^2. Also, any linear combination of elements in H that are orthogonal to G is also orthogonal to G. Hence,

U is a linear space. From Lemma 3.3 and the closure of H, it follows that any Cauchy sequence in H that is orthogonal to G converges to an element in H that is also orthogonal to G. Therefore, U is closed. ∎

Given a specification of G and using (3.7) as the definition of new information, one can always decompose a closed linear subspace of L^2. The Law of Iterated Projections can be proved using such a decomposition.

Lemma 3.5: Suppose that G and H are closed linear subspaces of L^2 and G is a subset of H. Then for any y in L^2, $P(y|G) = P[P(y|H)|G]$.

Proof: Using Lemma 3.4 we obtain an orthogonal decomposition of H into the sum of U given by (3.7) and G. Using Lemma 3.2 it follows that $P(y|H) = P(y|G) + P(y|U)$. Since $P(y|U)$ is in U and U is orthogonal to G, $P[P(y|U)|G] = 0$. Therefore, $P[P(y|H)|G] = P[P(y|G)|G] = P(y|G)$. ∎

Lemmas 3.2 and 3.5 offer an interesting comparison. Lemma 3.2 shows how to update a projection from a smaller information set G to larger information set H. In contrast, Lemma 3.5 shows how to calculate a projection onto a smaller information set G in terms of a projection onto a larger information set H.

As noted previously, one interpretation of the set G is that it contains past information. An alternative interpretation will also be used in some of our analysis. This interpretation involves a continuous linear functional π mapping H into the set of real numbers $\mathbf{R}$.

Definition 3.3: π is a linear functional on a linear subspace H of L^2 if π maps H into $\mathbf{R}$, and for any h_1 and h_2 in H and c_1 and c_2 in $\mathbf{R}$, $\pi(c_1 h_1 + c_2 h_2) = c_1\pi(h_1) + c_2\pi(h_2)$.

Definition 3.4: A linear functional π on a linear subspace H of L^2 is continuous if for any sequence $\{h_n : n \geq 1\}$ in H that converges to zero, the sequence of real numbers $\{\pi(h_n) : n \geq 1\}$ converges to zero.[3]

Linear functionals are important in studying competitive equilibrium pricing in environments with uncertainty. For instance, we can think of H as being a set of possible portfolio payoffs and π as a pricing function that assigns a price to each payoff. The pricing function can be used to construct G.

Lemma 3.6: Suppose H is a closed linear subspace of L^2 and π is a continuous linear functional on H. Let

$$G = \{g \in H : \pi(g) = 0\} . \tag{3.9}$$

Then G is a closed linear subspace of H.

Proof: The linearity of G follows from the linearity of π since for any g_1 and g_2 in G and any real numbers c_1 and c_2,

$$\begin{aligned} \pi(c_1 g_1 + c_2 g_2) &= c_1 \pi(g_1) + c_2 \pi(g_2) \\ &= 0 . \end{aligned} \tag{3.10}$$

The closure of G follows from the continuity of π. To see this, let $\{g_n : n \geq 1\}$ be a Cauchy sequence in G. Then $\{g_n : n \geq 1\}$ is a Cauchy sequence in H and hence converges to some element g_o in H. To see that g_o is also in G, notice that $\{g_n - g_0 : n \geq 1\}$ converges to zero. Since π is linear and continuous, $\{\pi(g_n) - \pi(g_o) \ : n \geq 1\}$ converges to zero. However, $\pi(g_n) = 0$ for all n so that $\pi(g_o) = 0$. Therefore, g_o is in G implying that G is closed. ▮

The set G given in (3.9) can be viewed as the set of all portfolio payoffs with zero prices. Taken together, Lemmas 3.4 and 3.6 provide an orthogonal decomposition of a closed linear subspace H of L^2. The next lemma provides a characterization of U as given by (3.7) when G is given by (3.9).

Lemma 3.7: Suppose H is a closed linear subspace of L^2 and π is a continuous linear functional on H. Then for G given by (3.9), the corresponding set U given by (3.7) satisfies

$$u = \pi(u)u^* \tag{3.11}$$

for all u in U and some u^* in U.

Proof: Consider two cases. First, suppose that U contains only the zero random variable. Then u^* can be chosen to be zero. Second, suppose that U contains some element u_o other than zero. Then $\pi(u_o)$ is different from zero since u_o is not in G. Let

$$u^* = u_o/\pi(u_o) . \tag{3.12}$$

For any u in U, $u - \pi(u)u^*$ is in G since $\pi(u^*)$ is one. Thus,

$$u = \pi(u)u^* \tag{3.13}$$

since $u - \pi(u)u^*$ is both in U and orthogonal to U and hence is orthogonal to itself. ▮

Lemma 3.7 shows that the set U is at most one-dimensional when a linear functional π is used to construct G. It turns out that an element in this one-dimensional set can be used to represent π.

Lemma 3.8: Suppose that H is a closed linear subspace of L^2 and that π is a continuous linear functional on H. Then there exists a unique element h^* in H such that $\pi(h) = E(hh^*)$.[4]

Proof: First, we study existence of h^*, and then we consider uniqueness of h^*. From Lemma 3.7 we know that there exists a u^* satisfying (3.11). Suppose u^* is zero. Then h^* can be chosen to be zero since $\pi(h)$ is zero for all h. Next suppose u^* is different from zero. Let $h^* = u^*/E(u^{*2})$. Notice that h^* is in U and

$$\pi(h^*) = 1/E(u^{*2}) \ . \tag{3.14}$$

Since $\pi(u^*)$ is one, for any h in H, $h - \pi(h)u^*$ is in G and hence orthogonal to h^*. Thus,

$$\begin{aligned} E(hh^*) &= \pi(h)E(u^*h^*) \\ &= \pi(h)E(u^{*2})/E(u^{*2}) \\ &= \pi(h) \ . \end{aligned} \tag{3.15}$$

Next, we show that h^* is unique. Let h^* and h^+ be any two elements in H for which $\pi(h) = E(hh^*) = E(hh^+)$ for all h in H. Then $h^* - h^+$ is orthogonal to H and in particular is orthogonal to itself. This can happen only if $h^* - h^+$ is zero. ∎

4. Infinite Sequences of Subspaces

In this section we study the limiting behavior of projections onto infinite sequences of closed linear subspaces of L^2. We present two types of results corresponding to whether the sequences are increasing or decreasing.

Definition 4.1: A sequence $\{H_n : n \geq 1\}$ of subsets of L^2 is increasing if $H_{n+1} \supset H_n$ for all $n \geq 1$.

Definition 4.2: A sequence $\{H_n : n \geq 1\}$ of subsets of L^2 is decreasing if $H_n \supset H_{n+1}$ for all $n \geq 1$.

First, we consider projections onto an increasing sequence $\{H_n : n \geq 1\}$ of closed linear subspaces of L^2. The set

$$\bigcup_{n \geq 1} H_n \tag{4.1}$$

is a linear subspace of L^2 that is not necessarily closed. We leave it as an exercise for the reader to verify that the set given in (4.1) is a linear subspace. Let H_o denote the closure of the linear subspace given in (4.1). The following result shows how to approximate projections onto H_o.

Lemma 4.1: Suppose $\{H_n : n \geq 1\}$ is an increasing sequence of closed linear subspaces of L^2. Then for any y in L^2, $\{P(y|H_n) : n \geq 1\}$ converges to $P(y|H_o)$.

Proof: Let y be any element of L^2. For $m \geq n$, H_m contains H_n. Consequently, Lemma 3.5 implies that $P(y|H_n) = P[P(y|H_m)|H_n]$ which in turn implies that $P(y|H_m) - P(y|H_n)$ is orthogonal to H_n. Therefore,

$$E\left\{[P(y|H_m) - P(y|H_n)]^2\right\} = E[P(y|H_m)^2] - E[P(y|H_n)^2] \;. \tag{4.2}$$

Since the left hand side of (4.2) is nonnegative, the sequence of real numbers $\{E[P(y|H_n)^2] : n \geq 1\}$ is nondecreasing. Furthermore,

$$E(y^2) = E[P(y|H_n)^2] + E\left\{[y - (y|H_n)]^2\right\} \geq E[P(y|H_n)^2] \;. \tag{4.3}$$

Thus, $\{E[P(y|H_n)^2] : n \geq 1\}$ is a bounded monotone sequence. Therefore, there exists a real number δ such that

$$\lim_{n\to\infty} E[P(y|H_n)^2] = \delta \;. \tag{4.4}$$

Limit (4.4) and equality (4.2) imply that

$$\begin{aligned} &\lim_{n\to\infty} \sup_{m\geq n} E\left\{[P(y|H_m) - P(y|H_n)]^2\right\} \\ &= \lim_{n\to\infty} \sup_{m\geq n} \left\{E[P(y|H_m)^2] - E[P(y|H_m)^2]\right\} \\ &= 0 \;. \end{aligned} \tag{4.5}$$

because $\{E[P(y|H_n)^2] : n \geq 1\}$ is a convergent and hence Cauchy sequence of real numbers. Thus, $\{P(y|H_n) : n \geq 1\}$ is Cauchy and hence converges to some element h_o in H_o.

To prove that $h_o = P(y|H_o)$, let h be any element in H_o. Then there exists a sequence $\{h_n : n \geq 1\}$ in H_o that converges to h for which h_n is in H_n. Thus by Lemma 3.3,

$$E[(y - h_o)h] = \lim_{n\to\infty} E\left\{[y - P(y|H_n)]h_n\right\} \;. \tag{4.6}$$

However, $E\{[y - P(y|H_n)]h_n\}$ is zero for all $n \geq 1$. Consequently,

$$E[(y - h_o)h] = 0 \tag{4.7}$$

for any h in H_o which proves that h_o solves the orthogonality problem (Problem 2.2). ∎

One way to construct an increasing sequence of closed linear subspaces of L^2 is as follows. Let $\{y_n : n \geq 1\}$ be a sequence in L^2, and let

$$H_n = \Big\{h : h = c_1 y_1 + c_1 y_2 + \cdots + c_n y_n \text{ for some real numbers } c_1, c_2, \ldots, c_n\Big\}. \tag{4.8}$$

Lemma 4.2: If $y_1, y_2, \ldots, y_n$ are in L^2, then H_n given by (4.8) is a closed linear subspace of L^2.

Proof: It is straightforward to prove that H_n is a linear subspace of L^2. We leave this as an exercise for the reader. To prove that H_n is closed, first suppose that $n = 1$. Let $\{h_m : m \geq 1\}$ be any Cauchy sequence in H_1. We consider two cases. First suppose y_1 is zero. Then h_m is zero for each m implying that $\{h_m : m \geq 1\}$ converges to zero. Hence H_1 is closed in this case. Next suppose y_1 is different from zero. Then for each $n \geq 1$, $h_m = c_m y_1$ for some real number c_m. Hence for $\ell \geq m$,

$$E[(h_\ell - h_m)^2] = (c_\ell - c_m)^2 E(y_1^2). \tag{4.9}$$

Since $\{h_n : n \geq 1\}$ is Cauchy in L^2 and $E(y_1^2) > 0$, $\{c_n : n \geq 1\}$ is a Cauchy sequence of real numbers. Thus $\{c_n : n \geq 1\}$ converges to some real number c_o. Let $h_o = c_o y_1$. Then

$$\begin{aligned} \lim_{n\to\infty} E[(h_n - h_o)^2] &= \lim_{n\to\infty} (c_n - c_o)^2 E(y_1^2) \\ &= 0 \end{aligned} \tag{4.10}$$

which proves that H_1 is closed since $c_o y_1$ is in H_1.

Next, suppose that H_{n-1} is closed. Let

$$U_n = \Big\{u : u = c[y_n - P(y_n|H_{n-1})] \text{ for some } c \text{ in } \mathbf{R}\Big\}. \tag{4.11}$$

Notice that U_n is a subset of H_n since elements of U_n are linear combinations of y_n and an element in H_{n-1}. By mimicking the proof that H_1 is closed, it can be shown that U_n is closed. Hence, Lemma 3.1 implies that $U_n + H_{n-1}$ is closed. Since H_{n-1} and U_n are subsets of H_n and

H_n is a subspace of L^2, $U_n + H_{n-1}$ is a subset of H_n. However, any element in H_n can be represented as

$$h = c_1 y_1 + c_2 y_2 + \cdots + c_n P(y_n|H_{n-1}) + c_n[y_n - P(y_n|H_{n-1})] \; . \tag{4.12}$$

Therefore, $U_n + H_{n-1} = H_n$ which proves that H_n is closed. ∎

Taken together, Lemmas 4.1 and 4.2 characterize the limiting behavior of projections as additional variables are included in the space onto which we are projecting. In particular, Lemma 4.1 shows that the sequence of projections converges to a random variable that can be interpreted as the projection onto the limiting space.

In some cases the increasing sequence of subspaces of L^2 is constructed from an underlying sequence of orthogonal spaces.

Definition 4.3: The sequence of subsets $\{U_n : n \geq 1\}$ of L^2 is orthogonal if U_n is orthogonal to U_m for all $n \geq 1$ and $m \geq 1$ for which $n \neq m$.

Definition 4.4: The sum of a sequence of orthogonal subsets of L^2 is

$$\sum_{n=1}^{\infty} U_n = \left\{ u : u = \sum_{n=1}^{\infty} u_n \text{ where } u_n \text{ is in } U_n \text{ and } \sum_{n=1}^{\infty} E(u_n)^2 < \infty \right\} .$$

A sequence $\{U_n : n \geq 1\}$ of orthogonal closed linear subspaces of L^2 can be used to construct an increasing sequence of closed linear subspaces of L^2. Let

$$H_1 = U_1 \; , \tag{4.13}$$

and

$$H_n = U_n + H_{n-1} \text{ for } n \geq 2 \; . \tag{4.14}$$

Then Lemma 3.1 guarantees that H_n is a closed linear subspace of L^2 for $n \geq 1$. Furthermore, since the zero random variable is in U_n, H_n contains H_{n-1}. The following result is the infinite dimensional counterpart to Lemmas 3.1 and 3.2.

Lemma 4.3: Suppose $\{U_n : n \geq 1\}$ is a sequence of orthogonal closed linear subspaces of L^2 and H_n is given by (4.13) and (4.14). Then

$$H_o = \sum_{n=1}^{\infty} U_n \tag{4.15}$$

and

(4.16) $$P(\cdot|H_0) = \sum_{n=1}^{\infty} P(\cdot|U_n) .$$

<u>Proof:</u> Let y be any element in L^2. By repeated application of Lemma 3.2, it follows that

(4.17) $$P(y|H_m) = \sum_{n=1}^{m} P(y|U_n) .$$

Lemma 4.1 implies that $\{P(y|H_m) : m \geq 1\}$ converges to $P(y|H_o)$ so that $\{P(y|H_m) : m \geq 1\}$ is a Cauchy sequence in H_o. Thus,

(4.18) $$\lim_{m\to\infty} \sup_{\ell\geq m+1} E\Big\{[\sum_{n=m+1}^{\ell} P(y|U_n)]^2\Big\} = 0 .$$

Since the sequence $\{U_n : n \geq 1\}$ is orthogonal,

(4.19) $$E\Big\{[\sum_{n=m+1}^{\ell} P(y|U_n)]^2\Big\} = \sum_{n=m+1}^{\ell} E[P(y|U_n)^2]$$

for any $\ell \geq m+1$. Thus (4.18) implies that

(4.20) $$\lim_{m\to\infty} \sum_{n=m+1}^{\infty} E[P(y|U_n)^2] = 0$$

which in turn implies that

(4.21) $$\sum_{n=1}^{\infty} E[P(y|U_n)^2] < \infty .$$

Consequently, $\sum_{n=1}^{\infty} P(y|U_n)$ is in $\sum_{n=1}^{\infty} U_n$ and is the limit point of $\{P(y|H_n) : n \geq 1\}$. This proves (4.16).

To prove (4.15), let h be any element of H_o. Then $h = P(h|H_o)$. Applying (4.16), it follows that

(4.22) $$h = \sum_{n=1}^{\infty} P(h|U_n) .$$

Hence, h is in $\sum_{n=1}^{\infty} U_n$. Conversely, suppose that u is in $\sum_{n=1}^{\infty} U_n$. Then u can be represented as

$$u = \sum_{n=1}^{\infty} u_n \tag{4.23}$$

where u_n is in U_n for $n \geq 1$. Notice that u_m is orthogonal to H_n for $m > n$. Hence, Lemma 3.3 guarantees that $\sum_{m=n+1}^{\infty} u_m$ is orthogonal to H_n. This in turn implies that

$$P(u|H_n) = \sum_{m=1}^{n} u_m \; . \tag{4.24}$$

Applying Lemma 4.1 then gives

$$\begin{aligned} P(u|H_o) &= \sum_{m=1}^{\infty} u_m \\ &= u \; . \end{aligned} \tag{4.25}$$

Therefore, u is in H_o. ∎

For the remainder of this section, we study decreasing sequences of subspaces. Let $\{H_n : n \geq 1\}$ be a decreasing sequences of subspaces of L^2. Define the limiting set to be

$$H_\infty = \bigcap_{n=1}^{\infty} H_n \; . \tag{4.26}$$

Lemma 4.4: Suppose $\{H_n : n \geq 1\}$ is a decreasing sequence of closed linear subspaces of L^2. Then H_∞ is a closed linear subspace of L^2.

Proof: To prove that H_∞ is a linear subspace of L^2, let h_1 and h_2 be any two elements in H_∞ and let c_1 and c_2 be any two real numbers. Then h_1 and h_2 are in H_n for all $n \geq 1$. Since H_n is a closed linear subspace of L^2, $c_1 h_1 + c_2 h_2$ is in H_n for all $n \geq 1$. Thus $c_1 h_1 + c_2 h_2$ is in H_∞ which proves that H_∞ is linear.

To prove that H_∞ is closed, let $\{h_n : n \geq 1\}$ be any Cauchy sequence in H_∞. Then $\{h_n : n \geq 1\}$ is in H_m for all $m \geq 1$. Since H_m is closed, $\{h_n : n \geq 1\}$ converges to some element h_∞ in H_m. Consequently, h_∞ is in H_∞ which proves that H_∞ is closed. ∎

Lemma 4.5: Suppose $\{H_n : n \geq 1\}$ is a decreasing sequence of closed linear subspaces of L^2. Then for any y in L^2, $\{P(y|H_n) : n \geq 1\}$ converges to $P(y|H_\infty)$.

Proof: Let y be any element of L^2. The first part of this proof mimics the proof of Lemma 4.1 except that for $m \geq n$, $P(y|H_m) = P[P(y|H_n)|H_m]$. Hence $P(y|H_n) - P(y|H_m)$ is orthogonal to $P(y|H_m)$. Consequently,

$$(4.27) \quad E\left\{[P(y|H_m) - P(y|H_n)]^2\right\} = E[P(y|H_n)^2] - E[P(y|H_m)^2] .$$

and $\{E[P(y|H_n)^2] : n \geq 1\}$ is a decreasing sequence with a lower bound of zero. Thus $\{P(y|H_m) : m \geq n\}$ is a Cauchy sequence in H_n for each $n \geq 1$. Since all of the H_n spaces are closed, the limit point h_∞ is in each of these spaces. Hence, h_∞ is also in H_∞.

To prove that $h_\infty = P(y|H_\infty)$, let h be any element of H_∞. Then for all $n \geq 1$, $y - P(y|H_n)$ is orthogonal to h since H_∞ is contained in H_n. Lemma 3.3 implies that

$$(4.28) \quad \begin{aligned} E[(y - h_\infty)h] &= \lim_{n\to\infty} E\left\{[y - P(y|H_n)]h\right\} \\ &= 0 . \end{aligned}$$

Therefore, h_∞ solves the orthogonality problem (Problem 2.2). ▮

When $\{H_n : n \geq 1\}$ is a decreasing sequence of closed linear subspaces of L^2, it is always possible to obtain an infinite dimensional orthogonal decomposition of any space in the sequence. To see this, let

$$(4.29) \quad U_n = \left\{u : u = h - P(h|H_{n+1}) \text{ for some } h \text{ in } H_n\right\} .$$

Lemma 3.4 shows that $H_n = U_n + H_{n+1}$ where U_n is a closed linear subspace of L^2 that is orthogonal to H_{n+1}. Also, for $m \geq n+1$, U_m is contained in H_{n+1} so that U_m is orthogonal to U_n. Consequently, $\{U_n : n \geq 1\}$ is an orthogonal sequence of closed linear subspaces of L^2.

Lemma 4.6: Suppose $\{H_n : n \geq 1\}$ is a decreasing sequence of closed linear subspaces of L^2. Then for any $n \geq 1$,

$$(4.30) \quad H_n = \sum_{m=0}^{\infty} U_{n+m} + H_\infty$$

where U_n is given by (4.29).

Proof: Given $n \geq 1$ and $\ell \geq 1$,

$$(4.31) \quad \begin{aligned} H_n &= U_n + H_{n+1} \\ &= U_n + U_{n+1} + \ldots + U_{n+\ell-1} + H_{n+\ell} . \end{aligned}$$

Notice that the sets on the right-hand side of (4.31) are mutually orthogonal. Let h be any element in H_n. Then applying Lemma 3.2,

$$h = P(h|H_n) = \sum_{m=0}^{\ell-1} P(h|U_{n+m}) + P(h|H_{n+\ell}) . \tag{4.32}$$

Lemma 4.5 implies that $\{h - P(h|H_{n+\ell}) : \ell \geq 1\}$ converges to $h - P(h|H_\infty)$. Consequently,

$$\left\{\sum_{m=1}^{\ell-1} P(h|U_{n+m}) : \ell \geq 1\right\} \tag{4.33}$$

converges. Lemma 4.3 guarantees that

$$\sum_{m=0}^{\infty} U_{n+m} \tag{4.34}$$

is a closed linear subspace of L^2. Hence, the random variable

$$\sum_{j=o}^{\infty} P(h|U_{n+j}) = h - P(h|H_\infty) \tag{4.35}$$

is in that closed linear space. Therefore,

$$h = \sum_{m=0}^{\infty} P(h|U_{n+m}) + P(h|H_\infty) \tag{4.36}$$

where $P(h|H_\infty)$ is in H_∞ and $\sum_{m=0}^{\infty} P(h|U_{n+m})$ is in $\sum_{m=o}^{\infty} U_{n+m}$. Relation (4.36) implies that the set on the left-hand side of (4.30) is contained in the set on the right-hand side. The equality follows from the fact that H_n is closed and U_{n+m} is contained in H_n for all m as is H_∞.
∎

Lemma 4.6 gives an important decomposition of decreasing sequences of closed linear subspaces of L^2. In the next two sections we will use this decomposition with two different notions of prediction. One of these notions of corresponds to calculating the conditional expectation and the other to finding the best linear prediction.

5. Conditional Expectations

In this section we show that conditional expectations can be viewed as a special case of projections. Our treatment of this topic presumes some familiarity with measure theory. Since this discussion is not essential to the subsequent analysis, readers may choose to proceed to section six.

Let (Ω, A, Pr) denote the underlying probability space, let x be a p-dimensional random vector defined on this space, let $\mathbf{R}^p$ denote the p-dimensional Euclidean space, and let

$$G = \{y \text{ in } L^2 : y = f(x) \text{ for some Borel measurable function } f \text{ mapping } \mathbf{R}^p \text{ into } \mathbf{R}\}\ . \tag{5.1}$$

Lemma 5.1: G given by (5.1) is a closed linear subspace of L^2.

Proof: Let B be the collection of subsets of Ω given by

$$B = \{b : b = \{\omega \text{ in } \Omega \text{ such that } x(\omega) \text{ is in } b^*\} \text{ for some Borel set } b^* \text{ in } \mathbf{R}^p\}\ . \tag{5.2}$$

We leave it as an exercise for the reader to verify that B is a subsigma algebra of A. In the appendix we prove that the L^2 space given in section two is complete. This space is defined using the sigma algebra A. By replacing A with B it follows that

$$G^* = \{y : y \text{ is measurable with respect to } B \text{ and } E(y^2) < \infty\} \tag{5.3}$$

is a closed linear subspace of L^2.

We will prove that $G = G^*$. First, suppose y is in G. Then y is in L^2 and $y = f(x)$ for some Borel measurable function f. Let b^+ be any Borel set in $\mathbf{R}$. Then

$$b^* = \{r : f(r) \text{ is in } b^+\} \tag{5.4}$$

is a Borel set in $\mathbf{R}^p$. Hence

$$\{\omega : f[x(\omega)] \text{ is in } b^*\} = \{\omega : x(\omega) \text{ is in } b^*\} \tag{5.5}$$

is in B. Therefore, y is in G^*.

Next, suppose y is in G^*. Then we can write $y = y^+ - y^-$ where

$$y^+ = \begin{cases} y & \text{if } y \geq 0 \\ 0 & \text{otherwise} \end{cases} \tag{5.6}$$

and

$$y^- = \begin{cases} -y & \text{if } y \leq 0 \\ 0 & \text{otherwise}\ . \end{cases} \tag{5.7}$$

Both y^+ and y^- are in G^*. Indicator functions for sets in B can be expressed as Borel measurable functions of x. Hence, simple functions (linear combinations of indicator functions) can also be expressed as Borel measurable functions of x. Since y^+ can always be approximated by an increasing sequence of simple functions, there exists an increasing sequence $\{f_n(x) : n \geq 1\}$ that converges almost surely to y^+ where f_n is a Borel measurable function for each n. Let

$$f^* = \lim_{n \to \infty} \sup f_n \tag{5.8}$$

and

$$f^+ = \begin{cases} f^* & \text{if } f^* < \infty \\ 0 & \text{otherwise} . \end{cases} \tag{5.9}$$

Then both f^* and f^+ are Borel measurable and $\{f_n(x) : n \geq 1\}$ converges almost surely to $f^+(x) = y^+$. A similar argument can be used to construct a Borel measurable function f^- such that $y^- = f^-(x)$. Let $f = f^+ - f^-$. Then f is Borel measurable and $y = f(x)$. ∎

Next we show that projections onto G are equal to expectations conditioned on x.

<u>Lemma 5.2:</u> For G given in (5.1), $P(z|G) = E(z|x)$.

<u>Proof:</u> Notice that B given in (5.2) is the smallest sigma algebra for which x is measurable. Let b be any set in B, and let b^* be the corresponding Borel set in $\mathbf{R}^p$ for which

$$b = \{\omega : x(\omega) \text{ is in } b^*\} . \tag{5.10}$$

We define

$$1_b = \begin{cases} 1 & \text{if } x \text{ is in } b^* \\ 0 & \text{otherwise} . \end{cases} \tag{5.11}$$

Then 1_b is in G since it is a Borel measurable function of x and $E(1_b^2) < \infty$. For any y in L^2, $P(y|G)$ solves the orthogonality problem (Problem 2.2). Hence

$$E\{[P(y|G) - y]1_b\} = 0 , \tag{5.12}$$

or equivalently

$$\int_b P(y|G)\, dPr = \int_b y\, dPr . \tag{5.13}$$

Since $P(y|G)$ is measurable with respect to B, and (5.13) holds for any b in B, it follows that $P(y|G) = E(y|x)$.[5] ▮

In light of this Lemma 5.2, we see that conditional expectations solve the least squares forecasting problem (Problem 2.1) as long as arbitrary (Borel measurable) nonlinear forecasting rules are permitted in constructing elements of G. Thus conditional expectations are special cases of projections.

Next, we shift from conditioning on a random vector to conditioning on current and past values of a vector stochastic process. Let $\{x(t) : -\infty < t < +\infty\}$ be a p-dimensional stochastic process, and let

(5.14)
$$H_n(t) = \Big\{ y \text{ in } L^2 : y = f[x(t),\ x(t-1),\ \ldots,\ x(t-n)] \text{ for some Borel measurable function } f \text{ mapping } \mathbf{R}^{p(n+1)} \text{ into } \mathbf{R} \Big\} .$$

Then Lemma 5.1 guarantees that $H_n(t)$ is a closed linear subspace of L^2, and Lemma 5.2 guarantees that projections onto $H_n(t)$ are the same as expectations conditioned on $x(t)$, $x(t-1)$, ..., $x(t-n)$. We define $H(t)$ to be the closure of

$$\bigcup_{n=1}^{\infty} H_n(t) . \tag{5.15}$$

Lemma 4.1 shows that projections onto $H(t)$ can be approximated by projections onto $H_n(t)$ for sufficiently large n. This raises the question of whether projections onto $H(t)$ can be interpreted as expectations conditioned on $\{x(t-n) :\ \text{for } n \geq 0\}$.

Lemma 5.3: $P[z|H(t)] = E[z|x(t),\ x(t-1),\ \ldots]$.

Proof: For each $n \geq 1$, let

(5.16)
$$B_n = \Big\{ b : b = \big\{ [x(t)',\ x(t-1)',\ \ldots,\ x(t-n)']' \text{ is in } b^* \big\} \text{ for some Borel set in } \mathbf{R}^{p(n+1)} \Big\} .$$

Then $\{B_n : n \geq 1\}$ is an increasing sequence of subsigma algebras of A. Let

$$B^* = \bigcup_{n \geq 1} B_n . \tag{5.17}$$

The collection of events B^* is an algebra since for any b_1 and b_2 in B^* the union of b_1 and b_2 is in B^*, and for any b in B^* the complement of b is in B^*. However, B^* is not necessarily a sigma algebra. For any b in B^*, the random variable

$$1_b = \begin{cases} 1 & \text{on } b \\ 0 & \text{otherwise} \end{cases} \tag{5.18}$$

is in $H_m(t)$ for some positive integer m. Hence, for any y in L^2, $y - P[y|H_n(t)]$ is orthogonal to 1_b for all $n \geq m$. It follows from Lemmas 3.3 and 4.1 that

$$\begin{aligned} E\Big\{\big\{y - P[y|H(t)]\big\}1_b\Big\} &= \lim_{n\to\infty} E\Big\{\big\{y - P[y|H_n(t)]\big\}1_b\Big\} \\ &= 0 \,. \end{aligned} \tag{5.19}$$

Therefore, $y - P[y|H(t)]$ is orthogonal to all indicator functions of sets in B^* or equivalently,

$$\int_b y\, dPr = \int_b P[y|H(t)]dPr \tag{5.20}$$

for all b in B^*.

As in the proof of Lemma 5.1, we write $y = y^+ - y^-$ where y^+ and y^- are given by (5.6) and (5.7) respectively. Since (5.20) holds for any y in L^2 it must hold for y^+ as well. For a given y^+, we define a measure μ on B^* to be

$$\mu(b) = \int_b y^+\, dPr = \int_b P[y^+|H(t)]\, dPr \,. \tag{5.21}$$

Let B_o be the smallest sigma algebra containing B^*. This sigma algebra is also the smallest sigma algebra for which $x(t)$, $x(t-1)$, ..., are measurable. There exists a unique extension of the measure μ from B^* to B_o.[6] Thus

$$\mu(b) = \int_b y^+\, dPr = \int_b P[y^+|H(t)]\, dPr \tag{5.22}$$

for all b in B_o. Recall that $\{P[y^+|H_n(t)] : n \geq 1\}$ converges to $P[y^+|H(t)]$. The Riesz-Fischer Theorem (see the appendix) implies that the space

$$G = \{y^* \text{ in } L^2 : y^* \text{ is measurable with respect to } B_o\} \tag{5.23}$$

is complete. Therefore, $P[y^+|H(t)]$ is measurable with respect to B_o and satisfies the defining characteristics for $E[y^+|x(t), x(t-1), \ldots]$.[7]

A similar argument can be applied to y^-. Combining these results and applying Lemma 2.6 gives

(5.24)
$$\begin{aligned} P[y|H(t)] &= P[y^+|H(t)] - P[y^-|H(t)] \\ &= E[y^+|x(t), x(t-1), \ldots] - E[y^-|x(t), x(t-1), \ldots] \\ &= E[y|x(t), x(t-1), \ldots] . \quad \blacksquare \end{aligned}$$

A convenient way to calculate projections is as follows. First obtain an orthogonal decomposition of the linear subspace being projected onto. Then calculate projections onto these smaller orthogonal subspaces and add them together. This approach can be used in calculating projections onto $H(t)$ since the sequence of subspaces $\{H(t-n) : n \geq 0\}$ is decreasing. Lemma 4.6 shows how to obtain an infinite-dimensional decomposition of this decreasing sequence. This decomposition uses the subspaces

(5.25)
$$U(t-n) = \left\{ u : u = h - P[h|H(t-1-n)] \text{ for some } h \text{ in } H(t-n) \right\}$$

for $n \geq 0$. In general, these subspaces will be infinite-dimensional. Also, projections onto $U(t)$ cannot be interpreted as conditional expectations although they can be interpreted as revisions in conditional expectations. In the next section, we consider an alternative notion of prediction that has the computational advantage that the space corresponding to $U(t-n)$ in (5.25) is finite-dimensional.

6. Linear Prediction Theory

In this section we consider an alternative notion of prediction to conditional expectations. This notion has the advantage that the associated orthogonal decompositions are easier to characterize. In order to obtain this simplification we restrict forecasting rules to be linear functions of an underlying set of random variables. As in section five, a sequence of closed linear spaces is generated using a p-dimensional stochastic process $\{x(t) : -\infty < t < +\infty\}$. We assume that each component of $x(t)$ is in L^2 for each time period t, and we define

(6.1)
$$\begin{aligned} H_n(t) = \{h : h = \alpha_1 \cdot x(t) + \alpha_2 \cdot x(t-1) + \ldots + \alpha_n \cdot x(t-n+1) \\ \text{for some vectors } \alpha_1, \alpha_2, \ldots, \alpha_n \text{ in } \mathbf{R}^p\} . \end{aligned}$$

Lemma 4.2 guarantees that $H_n(t)$ is a closed linear subspace of L^2 for each t and each $n \geq 1$. Also, $\{H_n(t) : n \geq 1\}$ is an increasing sequence of subspaces for each t. We define $H(t)$ to be the closure of

$$\bigcap_{n=1}^{\infty} H_n(t) . \tag{6.2}$$

Lemma 4.1 shows that projections onto $H(t)$ can be approximated by projections onto $H_n(t)$ for large values of n. We interpret the projection operator onto $H(t)$ as the best linear predictor given current and past values of $x(t)$.

One is tempted to think that any element in $H(t)$ can be represented as

$$\sum_{\tau=0}^{\infty} \alpha_\tau \cdot x(t-\tau) \tag{6.3}$$

for some sequence $\{\alpha_\tau : \tau \geq 0\}$ in $\mathbf{R}^p$. It turns out that this is not always true. For example, suppose that p is one, and let

$$x(t) = v(t) - v(t-1) \tag{6.4}$$

where

$$\begin{aligned} &E[v(t)^2] = 1 && \text{for all } t \\ &E[v(t)\,v(t-\tau)] = 0 && \text{for all } t \text{ and for all } \tau \neq 0 . \end{aligned} \tag{6.5}$$

First, we will show that $v(t)$ is in $H(t)$. To see this, let

$$y_n(t) = (n/n)\,x(t) + [(n-1)/n]\,x(t-1) + \ldots + (1/n)\,x(t-n+1) . \tag{6.6}$$

Then $y_n(t)$ is in $H(t)$ and

$$y_n(t) = v(t) - (1/n) \sum_{\tau=1}^{n} v(t-\tau) . \tag{6.7}$$

Notice that

$$E\Big\{[(1/n) \sum_{\tau=1}^{n} v(t-\tau)]^2\Big\} = (1/n) . \tag{6.8}$$

Therefore, $\{y_n(t) : n \geq 1\}$ converges to $v(t)$ proving that $v(t)$ is in $H(t)$. It turns out that $v(t)$ cannot be represented as

$$v(t) = \sum_{\tau=0}^{\infty} \alpha_\tau \cdot x(t-\tau) . \tag{6.9}$$

To see this, suppose to the contrary that (6.9) is true. Multiplying both sides of (6.9) by $v(t-j)$ and taking expectations gives

$$(6.10) \qquad \begin{aligned} 1 &= \alpha_0 && \text{for } j = 0 \\ 0 &= \alpha_j - \alpha_{j-1} && \text{for } j \geq 1 \ . \end{aligned}$$

These calculations use Lemma 3.3. Relation (6.10) implies that $\alpha_j = 1$ for all $j \geq 1$. However,

$$(6.11) \qquad \sum_{\tau=0}^{n} x(t-\tau) = v(t) - v(t-n-1) \ .$$

Thus,

$$(6.12) \qquad v(t) - \sum_{\tau=0}^{n} x(t-\tau) = v(t-n-1)$$

Clearly, the right-hand side of (6.12) does not converge to zero as n goes to infinity. Therefore, a representation of the form (6.3) does not exist for $v(t)$. Consequently, we cannot necessarily characterize the elements of $H(t)$ as infinite linear combinations of current and past values of $x(t)$.

It is often fruitful to characterize $H(t)$ in terms of an orthogonal decomposition. As in section five, we define

$$(6.13) \qquad U(t) = \Big\{u : u = h - P[h|H(t-1)] \text{ for some } h \text{ in } H(t)\Big\} \ .$$

The following result illustrates an advantage of using linear predictors.

Lemma 6.1: The set $U(t)$ given in (6.13) satisfies

$$(6.14) \qquad U(t) = \{u : u = \alpha \cdot u(t) \text{ for some } \alpha \text{ in } \mathbf{R}^p\}$$

where $u(t) = x(t) - P[x(t)|H(t-1)]$.[8]

Proof: Let $U^*(t)$ be the space defined by the right side of (6.14). Suppose h is in $H(t)$. We must prove that

$$(6.15) \qquad h - P[h|H(t-1)] = \alpha \cdot u(t)$$

for some α in $\mathbf{R}^p$. Since h is in $H(t)$, Lemma 4.1 implies that there exists a sequence $\{h_n : n \geq 1\}$ converging to h such that

$$(6.16) \qquad h_n = \alpha_0^n \cdot x(t) + \alpha_1^n \cdot x(t-1) + \ldots + \alpha_n^n \cdot x(t-n)$$

for some vectors $\alpha_0^n, \alpha_1^n, \ldots, \alpha_n^n$ in $\mathbf{R}^p$. Then

$$\begin{aligned} h_n - P[h_n|H(t-1)] &= \alpha_0^n \cdot \left\{x(t) - P[x(t)|H(t-1)]\right\} \\ &= \alpha_0^n \cdot u(t) \,. \end{aligned} \tag{6.17}$$

For $m \geq n$

$$\begin{aligned} E[(h_n - h_m)^2] = E\left\{[(\alpha_0^m - \alpha_0^n) \cdot u(t)]^2\right\} \\ + E\left\{P[(h_n - h_m)|H(t-1)]^2\right\} . \end{aligned} \tag{6.18}$$

Therefore, $\{\alpha_0^n \cdot u(t) : n \geq 1\}$ is a Cauchy sequence in $U^*(t)$ where $U^*(t)$ is given by the right-hand side of (6.14). Lemma 4.3 guarantees that $U^*(t)$ is closed so that there exists α in $\mathbf{R}^p$ such that $\{\alpha_0^n \cdot u(t) : n \geq 1\}$ converges to $\alpha \cdot u(t)$. Similarly, (6.18) implies that $\{P[h_n|H(t-1)] : n \geq 1\}$ converges to some random variable h_o in $H(t-1)$. For any h^* in $H(t-1)$,

$$\begin{aligned} E[(h - h_o)h^*] &= \lim_{n\to\infty} E\left\{\{h_n - P[h_n|H(t-1)]\}h^*\right\} \\ &= 0 \end{aligned} \tag{6.19}$$

so that $h_o = P[h|H(t-1)]$. Thus $\{h_n - P[h_n|H(t-1)] : n \geq 1\}$ converges to $h - P[h|H(t-1)] = \alpha \cdot u(t)$. ∎

Lemma 6.1 shows that $U(t)$ is a finite dimensional subspace of $H(t)$ containing linear combinations of the one-step ahead forecast errors in forecasting $x(t)$ using $H(t-1)$. This characterization underlies the Wold decomposition of a time series process. Let

$$H(-\infty) = \bigcap_{n=1}^{\infty} H(t-n) \,. \tag{6.20}$$

Notice that this definition of $H(-\infty)$ does not depend on the choice of time t. Lemma 4.6 shows that

$$H(t) = \sum_{n=0}^{\infty} U(t-n) + H(-\infty) \,. \tag{6.21}$$

Projecting $x(t)$ onto $H(t)$ and applying Lemma 4.6 gives

$$x(t) = \sum_{n=o}^{\infty} P[x(t)|U(t-n)] + P[x(t)|H(-\infty)] \,. \tag{6.22}$$

In light of Lemma 6.1, we obtain

$$x(t) = \sum_{n=0}^{\infty} \beta_n^t u(t-n) + P[x(t)|H(-\infty)] \tag{6.23}$$

where $\{\beta_n^t : n \geq 0\}$ is a sequence of $p \times p$ matrices of real numbers satisfying

$$\sum_{n=o}^{\infty} \text{trace}\left\{\beta_n^t E[u(t-n)u(t-n)']\beta_n^{t\prime}\right\} < \infty . \tag{6.24}$$

Representation (6.23) is referred to as the Wold decomposition of a stochastic process. Notice the coefficients of this representation depend on the time period t. In the special case that $\{x(t)\}$ is a covariance stationary stochastic process, the coefficients turn out to be time invariant. In this case, (6.23) is known as a "Wold representation for $x(t)$."

It is convenient to mention two (not exhaustive) categories of stochastic processes:

Definition 6.1: The stochastic process $\{x(t) : -\infty < t < +\infty\}$ is linearly regular if $H(-\infty) = \{0\}$.

Definition 6.2: The stochastic process $\{x(t) : -\infty < t < +\infty\}$ is linearly deterministic if $H(t) = H(-\infty)$ for all t.

When a stochastic process is linearly regular, the term $P[x(t)|H(-\infty)]$ drops out of the Wold decomposition. When a stochastic process is linearly deterministic,

$$x(t) = P[x(t)|H(t-n)] \tag{6.25}$$

for all n so that $x(t)$ can be forecast perfectly given past information. These two types of stochastic processes represent two extremes from the standpoint of linear prediction theory. It turns out that any stochastic process can be represented as the sum of a linearly regular process and a linearly deterministic process. Let

$$x^*(t) = x(t) - P[x(t)|H(-\infty)] , \tag{6.26}$$

and

$$x^+(t) = P[x(t)|H(-\infty)] . \tag{6.27}$$

Clearly,

$$x(t) = x^*(t) + x^+(t) . \tag{6.28}$$

We shall show that $\{x^*(t) : -\infty < t < +\infty\}$ is linearly regular and $\{x^+(t) : -\infty < t < +\infty\}$ is linearly deterministic.

Lemma 6.2: The process $\{x^*(t) : -\infty < t < +\infty\}$ is linearly regular.

Proof: Let $H^*(t)$ be defined in the same manner as $H(t)$ except with $x(t)$ replaced by $x^*(t)$. Also, let

$$H^*(-\infty) = \bigcap_{n=1}^{\infty} H^*(t-n) \,. \tag{6.29}$$

Now $x^*(t)$ is orthogonal to $H(-\infty)$ for all t which implies that $H^*(-\infty)$ is orthogonal to $H(-\infty)$. However, $x^*(t-n)$ is contained in $H(t)$ for all $n \geq 0$ so that $H^*(t)$ is contained in $H(t)$. Consequently, $H^*(-\infty)$ is contained in $H(-\infty)$. Therefore, $H^*(-\infty)$ is orthogonal to itself. The only subspace that is orthogonal to itself is $H^*(-\infty) = \{0\}$. ▮

Lemma 6.3: The process $\{x^+(t) : -\infty < t < +\infty\}$ is linearly deterministic.

Proof: Let $H^+(t)$ be defined in the same manner as $H(t)$ except with $x(t)$ replaced by $x^+(t)$. Also, let

$$H^+(-\infty) = \bigcap_{n=1}^{\infty} H^+(t-n) \,. \tag{6.30}$$

Now $x^+(t)$ is in $H(-\infty)$ for all t which implies that $H^+(t)$ is contained in $H(-\infty)$ for all t. Let h be any element of $H(-\infty)$. Then there exists a sequence $\{h_n : n \geq 1\}$ converging to h such that

$$h_n = \alpha_0^n \cdot x(t) + \alpha_1^n \cdot x(t-1) + \ldots + \alpha_n^n \cdot x(t-n) \tag{6.31}$$

for some vectors $\alpha_0^n, \alpha_1^n, \ldots, \alpha_n^n$ in $\mathbf{R}^p$. Note that

$$E[(h_n - h)^2] \geq E\big\{P[h_n - h|H(-\infty)]^2\big\} \tag{6.32}$$

so that $\{P[h_n|H(-\infty)] : n \geq 1\}$ converges to $P[h|H(-\infty)] = h$. However, $P[h_n|H(-\infty)]$ is in $H^+(t)$ which proves that $H^+(t)$ contains $H(-\infty)$. Therefore $H^+(t) = H(-\infty) = H^+(-\infty)$ for all t. ▮

In many applications in economics, including all of those contained in this book attention is restricted to processes that are linearly regular. In making such a restriction, we have a decomposition such as (6.28) in mind. Then our attention is focused on the portion of the forecasting problems that can be related to the one-step ahead forecast errors in current and past values of $x(t)$. This portion of the prediction problems is dynamic in the sense that information is accumulating over time.

Appendix

In this appendix we prove that the space L^2 is complete. This is a special case of the Riesz-Fischer Theorem. Our proof of this theorem uses the Monotone and Dominated Convergence Theorems.

Theorem A.1: Suppose $\{y_n : n \geq 1\}$ is a Cauchy sequence in L^2. Then there exists a random variable y_o in L^2 such that $\{y_n : n \geq 1\}$ converges to y_o.

Proof: Since $\{y_n : n \geq 1\}$ is Cauchy, for each $k \geq 1$ there is an n_k such that

$$(A.1) \qquad E[(y_m - y_{n_k})^2]^{\frac{1}{2}} \leq 1/2^k \text{ for all } m \geq n_k .$$

Therefore,

$$(A.2) \qquad \sum_{j=1}^{\infty} E[(y_{n_{j+1}} - y_{n_j})^2]^{\frac{1}{2}} \leq 1 .$$

By the Triangle Inequality,

$$(A.3) \qquad E\left\{\left[\sum_{j=1}^{k} |y_{n_{j+1}} - y_{n_j}|\right]^2\right\}^{\frac{1}{2}} \leq 1 \qquad \text{for all } k \geq 1 .$$

For $k \geq 1$, let

$$(A.4) \qquad y_k^* = |y_{n_1}| + \sum_{j=1}^{k} |y_{n_{j+1}} - y_{n_j}| .$$

The sequence $\{y_k^* : k \geq 1\}$ is nondecreasing. Consequently, y_o^* given by

$$(A.5) \qquad y_o^* = \lim_{k \to \infty} y_k^*$$

is a well-defined random variable that is possibly infinite-valued. By the Triangle Inequality and (A.3),

$$(A.6) \qquad [E(y_k^*)^2]^{\frac{1}{2}} \leq E[(y_{n_1})^2]^{\frac{1}{2}} + 1 ,$$

or equivalently,

$$E[(y_k^*)^2] \leq \left\{E[(y_{n_1})^2]^{\frac{1}{2}} + 1\right\}^2 \text{ for all } k \geq 1 . \tag{A.7}$$

Thus, by the Monotone Convergence Theorem,

$$\begin{aligned} E[(y_o^{*2})] &= \lim_{k\to\infty} E[(y_k^*)^2] \\ &\leq \left\{E[(y_{n_1})^2]^{\frac{1}{2}} + 1\right\}^2 \end{aligned} \tag{A.8}$$

so that y_o^* is finite with probability one and is in L^2.

Now,

$$y_{n_k} = y_{n_1} + \sum_{j=1}^{k} (y_{n_{j+1}} - y_{n_j}) . \tag{A.9}$$

Also, absolutely summable sequences of real numbers are summable. Thus, y_o given by

$$y_o = \begin{cases} \lim_{k\to\infty} y_{n_k} & \text{if } y_o^* \text{ is finite} \\ 0 & \text{otherwise .} \end{cases} \tag{A.10}$$

is a well-defined random variable since y_o^* is a random variable. Equalities (A.4) and (A.5) imply that

$$|y_{n_k}| \leq y_o^* \text{ for all } k \geq 1 , \tag{A.11}$$

which in turn implies that

$$|y_o| \leq y_o^* . \tag{A.12}$$

Consequently, y_o is in L^2. Inequalities (A.11) and (A.12) and the Triangle Inequality guarantee that

$$(y_o - y_{n_k})^2 \leq 4(y_o^*)^2 . \tag{A.13}$$

It follows from the Dominated Convergence Theorem that

$$\lim_{k\to\infty} E[(y_o - y_{n_k})^2] = 0 \tag{A.14}$$

since

$$\lim_{k\to\infty} y_{n_k} = y_o \tag{A.15}$$

on a set of probability measure one. Thus, we have proved that $\{y_{n_k} : k \geq 1\}$ converges to y_o in L^2.

To complete this proof, we must show that $\{y_n : n \geq 1\}$ converges to y_o in L^2. Given $\delta > 0$, there is k such that

$$E[(y_m - y_{n_k})^2]^{\frac{1}{2}} < \delta/2 \tag{A.16}$$

for all $m \geq n_k$ and

$$E[(y_{n_k} - y_o)^2]^{\frac{1}{2}} < \delta/2 \, . \tag{A.17}$$

By the Triangle Inequality,

$$E[(y_m - y_o)^2]^{\frac{1}{2}} < \delta \text{ for } m \geq n_k \, . \quad \blacksquare \tag{A.18}$$

Notes

1. We follow the usual convention of viewing the equivalence class of random variables that are equal almost surely as a unique random variable.

2. This notion of convergence is induced by the norm, $||y|| = E(y^2)^{\frac{1}{2}}$ on the linear space L^2.

3. The definition of continuity given here only considers the behavior of the function π in the neighborhood of zero. However, since c is linear, continuity at a single point immediately translates into continuity at all points.

4. This lemma is a special case of the Riesz Representation Theorem.

5. We take the notation $E(z|x)$ to mean the expectation conditioned on the sigma algebra generated by x.

6. The unique extension of measures from algebras to sigma algebras follows from the Carathéodory Theorem. See Halmos (1950) or Royden (1968) for a discussion of this result.

7. Recall that a projection is defined only up to an equivalence class of random variables that are almost surely equal. Some (but not necessarily all) members of this equivalence class will be measurable with respect to B_o.

8. By $P[x(t)|H(t-1)]$ we mean the vector containing the projections of components $x(t)$ onto $H(t-1)$.

3

Exact Linear Rational Expectations Models: Specification and Estimation

by Lars Peter HANSEN and Thomas J. SARGENT

Introduction

A distinguishing characteristic of econometric models that incorporate rational expectations is the presence of restrictions across the parameters of different equations. These restrictions emerge because people's decisions are supposed to depend on the stochastic environment which they confront. Consequently, equations describing variables affected by people's decisions inherit parameters from the equations that describe the environment. As it turns out, even for models that are linear in the variables, these cross-equation restrictions on the parameters can be complicated and often highly nonlinear.

This paper proposes a method for conveniently characterizing cross-equation restrictions in a class of linear rational expectations models, and also indicates how to estimate statistical representations satisfying these restrictions. For most of the paper, we restrict ourselves to models in which there is an *exact* linear relation across forecasts of future values of one set of variables and current and past values of some other set of variables. The key requirement is that all of the variables entering this relation must be observed by the econometrician. While probably only a minority of rational expectations models belong to this class, it does contain interesting models that have been advanced to study the term structure of interest rates, stock prices, consumption and permanent income, the dynamic demand for factors of production, and many other subjects.

It is useful to compare the class of exact models with the class studied by Hansen and Sargent (1980a). The differences lie entirely in the interpretations of the *error terms* in the equations that are permitted. In Hansen and Sargent (1980a), random processes which the econometrician treats as disturbances in decision rules can have a variety of sources. Disturbance terms can be interpreted as reflecting shocks to technologies or preferences observed by private agents but not by the econometrician. Disturbances can also be interpreted as reflecting interactions with *hidden* decision variables which are simultaneously chosen by private agents but unobserved by the econometrician. Finally, disturbances can be interpreted, along the lines of Shiller (1972), as reflecting the phenomenon that in forecasting the future, private agents use larger information sets than the econometrician can consider because of data limitations. Of these alternative interpretations of error terms, only the last one can be accommodated within the class of exact models of the present paper. While this limitation on the permissible interpretations of error terms excludes many rational expectations models, a variety of interesting examples still remains within the general class of exact linear rational expectations models.

In linear rational expectations models, the cross-equation restrictions can be characterized very conveniently by working in terms of a vector moving-average representation for the variables being modeled. By straightforward applications of the Wiener-Kolmogorov least squares prediction formulas, these restrictions can readily be deduced. Once the restrictions are deduced, the parameters of the model can be estimated by maximizing one of various approximations to the likelihood function. The vector moving-average representation that incorporates the rational expectations restrictions is nested within less constrained vector moving-average representations, and a likelihood ratio statistic can be computed to test the model.

The ease of characterizing the restrictions and calculating estimates is a great virtue of specifying the model in vector moving-average form. However, an identification question must be addressed before this strategy can be implemented. Without *a priori* restrictions on their parameters, many vector moving-average representations are consistent with a given set of second moments. A natural and practically important question is whether the cross-equation rational expectations restrictions provide enough prior information to identify a unique moving-average representation. For the case of exact linear rational expectations models, Section 3 provides results that characterize identification. Insofar as the identification question is concerned, there are substantial differences between exact rational expectations models and models that admit one or more of the additional interpretations of the error terms described above. It is the special nature of the identification problem in these exact models, and not anything special about the appropriate methods either of representing the models or of estimating them, that causes us to restrict this paper mainly to analyzing exact linear rational expectations models. In Section 6, we briefly indicate how both our methods for model specification and estimation carry over to *inexact* linear rational expectations models.

As Shiller (1979) and Hansen and Hodrick (1980) have indicated for several special examples of our general model, it is possible to devise powerful tests of such models without estimating the complete vector process subject to the model's restrictions. However, for many applications, the analyst wants more than just a test of the model, and desires a complete representation of the vector process. Indeed, our interest in the identification and estimation of constrained moving-average models is not entirely motivated by the exact linear rational expectations models that occupy most of our attention in the present paper. As we indicate in Section 6, the restrictions that emerge in the present models strongly resemble those that characterize rational expectations models which can accommodate additional interpretations of disturbance terms (e.g. Hansen and Sargent 1980a). This makes constrained moving-average estimation a more generally useful method for estimating the parameters needed to overcome Lucas's (1976) critique of econometric policy evaluation procedures.

1. General Model

In this section we specify a general time series econometric model and consider representations of solutions to that model. We begin by specifying in turn the information of economic agents, the information set of the econometrician, and the economic model.

Economic Agents' Information Set

Let J^+ denote the set of nonnegative integers, J the set of all integers, and $x \equiv \{x_t : t \in J\}$ be a p-dimensional, vector stochastic process that is covariance stationary and has finite second moments. For simplicity, we assume that $Ex_t = 0$. The time t common information set of economic agents, denoted Ω_t, is the set of all random variables with finite second moments that are (Borel measurable) functions of current and past values of x. In making decisions at time t, economic agents are assumed to forecast optimally conditioned on Ω_t. Let $\mathrm{P}(\cdot|\Omega_t)$ denote the projection operator that maps random variables with finite second moments into optimal forecasts based on Ω_t. This operator can be interpreted as an expectation operator conditioned on current and past values of x.

Econometrician's Information Set

The econometrician is assumed to observe only the first q components of the x process. We denote the resulting process $y \equiv \{y_t : t \in J\}$. Let Σ_t denote the closed (in mean-square) information set generated by taking linear combinations of current and past values of y. Instead of forecasting using conditional expectations, the econometrician uses the linear least squares projections onto Σ_t. In general, forecast errors using $\mathrm{P}(\cdot|\Omega_t)$ will have smaller second moments than forecast errors using $\mathrm{P}(\cdot|\Sigma_t)$.

We assume that y is linearly regular and has full rank. These assumptions are sufficient to guarantee that elements of Σ_t can be represented as

$$\sum_{j=0}^{\infty} \alpha_j \cdot v_{t-j} \tag{1.1}$$

where $v \equiv \{v_t : t \in J\}$ is a serially uncorrelated q-dimensional process satisfying $Ev_t = 0$, $Ev_t v_t' = I$ where the entries of v_t are in Σ_t. Also, $\alpha \equiv \{\alpha_j : j \in J\}$ is a sequence in $\mathbf{R}^q$ satisfying

$$\sum_{j=0}^{\infty} \alpha_j \cdot \alpha_j < \infty \,. \tag{1.2}$$

The v process is said to be *fundamental* for the y process and v_t is the new uncorrelated information that is added to Σ_{t-1} in forming Σ_t. Since $P(v_{t+1}|\Sigma_t) = 0$, it is more convenient to represent Σ_t in terms of current and past values of v instead of current and past values of y.

We assume that the econometrician uses a subset of the information observed by economic agents and that he calculates linear least squares projections in making these forecasts. These are convenient assumptions for studying many dynamic economic models using time series methods.

Economic Model

Our general model is of the form:

$$P\Big\{[A(L) \; ; \; B(L^{-1})]y_t|\Omega_{t-\ell}\Big\} = 0 \tag{1.3}$$

where

$$A(z) = A_n(z)/A_d(z) \,, \tag{1.4}$$

"driving process"
Euler eqn or Expectations eqn.

$$B(z) = B_n(z)/B_d(z)\,, \tag{1.5}$$

$A_n(z)$ is an $(r \times r)$ matrix polynomial with a determinant that has zeros outside the unit circle of the complex plane, $B_n(z)$ is an $[r \times (q-r)]$ matrix polynomial, and $A_d(z)$ and $B_d(z)$ are scalar polynomials with zeros outside the unit circle of the complex plane.

Solution

Let Λ_t be generated by current and past values of a serially uncorrelated, q-dimensional process $w \equiv \{w_t : t \in J\}$ where $Ew_t = 0$ and $Ew_t w_t' = I$. Clearly, $P(w_{t+1}|\Lambda_t) = 0$ and the entries of w_t are in Λ_t. We assume that solutions to the model can be represented as time invariant linear functions of current and past values of w:

$$y_t = C(L)w_t = \begin{bmatrix} C_1(L) \\ C_2(L) \end{bmatrix} w_t \tag{1.6}$$

where

$$C(z) \equiv \sum_{j=0}^{\infty} c_j z^j \quad \text{and} \quad \sum_{j=0}^{\infty} \operatorname{trace}(c_j c_j') < \infty\,. \tag{1.7}$$

Partition C_1 is $(r \times q)$ and partition C_2 is $[(q-r) \times q]$.

One possible candidate for w is $w = v$, in which case $\Sigma_t = \Lambda_t$ for all t. We do not restrict ourselves to this specification and instead allow Λ_t to be strictly larger than Σ_t. For the purposes of solving the model, we take Λ_t to be a subset of Ω_t. Since some of the components of Ω_t may not be observed by the econometrician, it is possible for Λ_t to be strictly larger than Σ_t.

Applying the Law of Iterated Projections to (1.3) gives

$$P\Big\{[A(L)\ ;\ B(L^{-1})]y_t|\Lambda_{t-\ell}\Big\} = 0 \tag{1.8}$$

or equivalently

$$P\Big\{[A(L)\ ;\ B(L^{-1})]y_t|\Lambda_t\Big\} = D_1(L)w_t \tag{1.9}$$

where $D_1(z)$ is an $(r \times q)$ matrix polynomial with degree $\ell - 1$. When $\ell = 0$, $D_1(z) = 0$. Also, (1.6) implies that

$$[0\ ;\ I]\, y_t = D_2(L)w_t \tag{1.10}$$

where $D_2(z) = C_2(z)$. Our goal is to solve for $C(z)$ in terms of the triple $[A(z), B(z), D(z)]$ where

$$D(z) = \begin{bmatrix} D_1(z) \\ D_2(z) \end{bmatrix} . \tag{1.11}$$

Relation (1.9) implies that

$$\left\{[A(z) \; ; \; B(z^{-1})] \begin{bmatrix} C_1(z) \\ C_2(z) \end{bmatrix}\right\}_+ = D_1(z) , \tag{1.12}$$

or equivalently

$$A(z)C_1(z) + [B(z^{-1})C_2(z)]_+ = D_1(z) \tag{1.13}$$

where $[\cdot]_+$ is the annihilation operator that is defined as follows. The matrix $B(z^{-1})C_2(z)$ is well defined for $\rho < |z| < 1$ for some $\rho < 1$. In particular, ρ can be chosen so that ρ^{-1} is the modulus of the smallest zero of $B_d(z)$. The matrix function $B(z^{-1})C_2(z)$ has a two-sided Laurent series expansion in the region $\{z : \rho < |z| < 1\}$. Then $[B(z^{-1})C_2(z)]_+$ is the function defined by the power series expansion where the negative powers of z are ignored. Negative powers of z correspond to future values of w_t. Hence, the projection of these terms onto Σ_t is zero. Solving for $C_1(z)$ gives

$$\begin{aligned} C_1(z) &= A(z)^{-1}\left\{D_1(z) - [B(z^{-1})D_2(z)]_+\right\} \\ C_2(z) &= D_2(z) . \end{aligned} \tag{1.14}$$

Equation (1.14) can be viewed as linear transformations mapping the matrix function $D(z)$ into the matrix function $C(z)$. Throughout this paper we will think of alternative solutions to (1.8) as being indexed by alternative choices of $D(z)$. The only restriction we have placed on $D(z)$ is that $D_1(z)$ be a polynomial with degree $\ell - 1$, and that $D_2(z)$ have a power series expansion with square summable coefficients.

To obtain an alternative convenient representation of $C(z)$, we follow Hansen and Sargent (1980a) by characterizing the annihilation operator as applied to the function $B(z^{-1})D_2(z)$ in terms of properties of this function inside the unit disk of the complex plane. The functions $D_1(z)$ and $D_2(z)$ are both analytic inside the unit disk, and the function $B(z^{-1})$ is analytic everywhere in this same region except at the zeros of $B_d(z^{-1})$ and possibly at $z = 0$. The zeros of $B_d(z)$ are all outside the unit circle of the complex plane, implying that the zeros of $B_d(z^{-1})$ are all inside the unit circle. Since $B_d(z)$ is a finite-order polynomial, $B_d(z^{-1})$ has a finite number of zeros. Hence $B(z^{-1})D_2(z)$ is analytic everywhere inside the unit disk except at a finite number of points. At each of the points for which it is not analytic, $B(z^{-1})D_2(z)$ has a finite-order pole.

To compute $[B(z^{-1})D_2(z)]_+$, we take a *partial fractions decomposition* whereby we form the principal parts of the Laurent series expansion about each of the poles and subtract the sum of these principal parts from $B(z^{-1})D_2(z)$. Let z_k denote the kth pole and $G_k(z^{-1})$ denote the principal part of the Laurent series expansion of $B(z^{-1})D_2(z)$ about z_k. Recall that the principal part, $G_k(z^{-1})$, of a Laurent series expansion about z_k consists only of the terms $(z-z_k)^j$ for strictly negative powers of j. Since z_k is a finite-order pole, $G_k(z^{-1})$ is the sum of only a finite number of such terms. Then $B(z^{-1})D_2(z) - G(z^{-1})$ is analytic inside the unit circle of the complex plane where

$$G(z^{-1}) \equiv \sum_k G_k(z^{-1}) \,. \tag{1.15}$$

In other words, $B(z^{-1})D_2(z) - G(z^{-1})$ has a power series expansion that is convergent in the unit disk of the complex plane.

By construction, $G(z)$ satisfies the following restriction:

Restriction R1: $G(z)$ is an $(r \times q)$ function that is analytic in an open set containing $\{z : |z| \leq 1\}$ and $G(0) = 0$.

For any function G that satisfies *R1*, G has a power series expansion with a leading coefficient zero and a radius of convergence that exceeds one. Consequently, in a domain containing $\{z : |z| \geq 1\}$, $G(z^{-1})$ can be represented as

$$G(z^{-1}) = \sum_{j=1}^{\infty} g_j z^{-j} \,. \tag{1.16}$$

Since $B(z^{-1})D_2(z) - G(z^{-1})$ is analytic in the unit disk and $G(z)$ satisfies *R1*,

$$[B(z^{-1})D_2(z)]_+ = B(z^{-1})D_2(z) - G(z^{-1}) \,. \tag{1.17}$$

In summary, one convenient way to compute the left side of (1.14) is first to compute $G(z^{-1})$ and then subtract it from $D_1(z) - B(z^{-1})D_2(z)$. Substituting into (1.14), we have that

$$\begin{aligned} C_1(z) &= A(z)^{-1}[D_1(z) - B(z^{-1})D_2(z) + G(z^{-1})] \\ C_2(z) &= D_2(z) \,. \end{aligned} \tag{1.18}$$

Equation (1.18) implies that

$$[A(z);\ B(z^{-1})]\,C(z) = z^{\ell} H(z^{-1})\ , \tag{1.19}$$

where

$$H(z^{-1}) \equiv z^{-\ell}\,[D_1(z) + G(z^{-1})]\ . \tag{1.20}$$

Then $H(z)$ satisfies *R1* by construction. Furthermore, for any $C(z)$ satisfying (1.19) for some $H(z)$ that satisfies *R1*, there exists an *admissible* $D(z)$ such that (1.14) is satisfied where an admissible $D(z)$ is one with square summable coefficients and an upper $(r \times q)$ partition that is a matrix polynomial with degree $\ell - 1$. In accordance with relation (1.20),

$$D_1(z) = \sum_{j=1}^{\ell} d_j\, z^j\ , \tag{1.21}$$

and in accordance with (1.14),

$$D_2(z) = C_2(z)\ . \tag{1.22}$$

Therefore, an equivalent characterization to $C(z)$ satisfying (1.14) for some admissible $D(z)$ is that $C(z)$ satisfies (1.19) for some $H(z)$ satisfying *R1*. Characterization (1.14) has the advantage that the solution is explicitly parameterized by $D(z)$. However, characterization (1.19) also will turn out to also be useful in Section 3 when we investigate issues pertaining to identification of $D(z)$ and hence $C(z)$.

In checking (1.19), it suffices to verify that $H(z)$ is analytic in the region $\{z : |z| < 1\}$ rather than in an open set containing $\{z : |z| \leq 1\}$ because the left side of (1.19) is analytic in the region $\{z : \rho < |z| < 1\}$. Consequently, when $H(z)$ is analytic inside the unit circle of the complex plane, it is also analytic in the larger set $\{z : |z| < (1/\rho)\}$.

2. Examples

In this section we consider five illustrations of the general model specified and solved in section one.

Example 1: Lognormal Model of Bond Pricing

An implication of the intertemporal asset pricing model as studied by LeRoy (1973), Rubinstein (1976), Lucas (1978), Breeden (1979), and Cox, Ingersoll and Ross (1985), among others, is that the price of a pure discount bond that pays one dollar n time periods into the future satisfies:

$$\exp(p_t^n) = \mathrm{P}\left[\exp(m_{t+n} - m_t)|\Omega_t\right] \tag{2.1}$$

where $\exp(p_t^n)$ is the bond price, $\exp(m_t)$ is the indirect marginal utility of money, and Ω_t is the information set available to economic agents, all at date t. Following Hansen and Singleton (1983), Hansen and Hodrick (1983), Campbell (1986), Harvey (1988), and Hansen and Singleton (1990), suppose that $m_t - m_{t-1}$ is one component of x_t where x is a stationary Gaussian process.[1] Then (2.1) implies that

$$\begin{aligned} p_t^n &= \mathrm{P}\left[(m_{t+n} - m_t)|\Omega_t\right] + \mathbf{c}_n \\ &= \mathrm{P}\left[(m_{t+1} - m_t) + (m_{t+2} - m_{t+1}) + \ldots \right. \\ &\quad \left. + (m_{t+n} - m_{t+n-1})|\Omega_t\right] + \mathbf{c}_n \end{aligned} \tag{2.2}$$

where $\mathbf{c}_n$ is equal to one-half the conditional variance of $m_{t+n} - m_t$. Abstracting from the constant term, relation (2.2) is a special case of the general model described in section one with $y_{1t} = p_t^n$; the first entry of y_{2t} equal to $m_t - m_{t-1}$ and the remaining entries being variables that are observed by the econometrician and are potentially useful in forecasting future values of $m_t - m_{t-1}$; $A(z) = 1$, $B(z) = (z + z^2 + \ldots + z^n)\,[1;\ 0]$, and $\ell = 0$. Therefore, $C(z)$ satisfies

$$\begin{aligned} C_1(z) &= -[1;\ 0]\left[(z^{-1} + z^{-2} + \ldots + z^{-n})\, D_2(z)\right]_+ \\ C_2(z) &= D_2(z)\ . \end{aligned} \tag{2.3}$$

Next suppose that the indirect marginal utility for money is not observable to an econometrician. Instead observations are available on the price of a one-period discount bond. Setting $n = 1$ in (2.2) results in

$$p_t^1 = \mathrm{P}\left[(m_{t+1} - m_t)|\Omega_t\right] + \mathbf{c}_1\ . \tag{2.4}$$

Applying the Law of Iterated Projections (see Lemma 3.5 in Chapter 2) to (2.2) and substituting from (2.4), it follows that

$$p_t^n = \mathrm{P}\left[(p_t^1 + p_{t+1}^1 + \ldots + p_{t+n-1}^1)|\Omega_t\right] + \mathbf{c}_n - n\mathbf{c}_1\ . \tag{2.5}$$

Again abstracting from the constant term, relation (2.5) is also in the form of the general model presented in section one where now the first entry of y_{2t} is p_t^1 and $B(z) = (1 + z + \ldots + z^{n-1})\,[1;\ 0]$.

Finally, replacing $-p_t^n/n$ in (2.5) by the one-period rate of return on an n-period discount bond and p_t^1 by the one-period rate of return on a one-period bond, one obtains Sargent's (1979) version of a rational expectations model of the term structure of interest rates.

Example 2: Present-Value Model

Let d_t denote the time t dividend or payout, and p_t the time t value to owning a claim to this dividend process from time t forward. Suppose these two variables are related via a present-value formula

$$p_t = \mathrm{P}\,(\sum_{j=0}^{\infty} \lambda^\tau\, d_{t+\tau} \mid \Omega_t)\,, \tag{2.6}$$

where λ is a constant discount factor.[2] In LeRoy and Porter (1981) and Shiller (1981), d_t is the dividend paid to share-holders of a stock and p_t is the price of the stock. In Hamilton and Flavin (1986) and Roberds (Chapter 6), d_t is the net surplus of a government (taxes minus expenditures) and p_t is government debt.

As noted in Campbell and Shiller (1987), a present-value model of the form (2.6) is a special case of an exact linear rational expectations model as defined in section one. To see this, let $y_{1t} = p_t$, d_t be the first entry of y_{2t}, $A(z) = 1$, $B(z) = -1/(1-\lambda z)\,[1;\,0]$ and $\ell = 0$. The solution to the model is

$$\begin{aligned} C_1(z) &= [1;\,0]\,[zD_2(z) - \lambda D_2(\lambda)]/(z-\lambda) \\ C_2(z) &= D_2(z)\,. \end{aligned} \tag{2.7}$$

Example 3: Permanent Income Model of Consumption

Following Hall (1978), Flavin (1981), Hansen (1987) and Sargent (1987b), we consider a rational expectations version of the permanent income model of the form:

$$c_t - \rho a_t = (1-\lambda)\,\mathrm{P}\,(\sum_{j=0}^{\infty} \lambda^j\, e_{t+j} \mid \Omega_t) \tag{2.8}$$

where a_t is the asset-holding, c_t is consumption and e_t is the endowment or labor income of a consumer at time t. The parameters ρ and λ are related via $\lambda = 1/(1+\rho)$. This model has a built-in stochastic singularity because of the resource constraint:

$$c_t + a_{t+1} - a_t = \rho a_t + e_t\,. \tag{2.9}$$

For this reason, instead of treating a, c and e as distinct processes, we investigate implications for $c_t - \rho a_t$ and e_t. Expressed in terms of these transformed process, the model in equation (2.8) is essentially the same as the present-value model.

The permanent income model is known to have the property that $c_{t+1} - c_t$ is uncorrelated with y_{t-j} for $j \in J^+$. To verify this property, note that

$$
\begin{aligned}
c_{t+1} - c_t &= (c_{t+1} - \rho a_{t+1}) - (c_t - \rho a_t) + \rho(a_{t+1} - a_t) \\
&= (c_{t+1} - \rho a_{t+1}) - (c_t - \rho a_t) - \rho(c_t - \rho a_t) + \rho e_t \, . \\
&= (L^{-1} - \lambda^{-1})(c_t - \rho a_t) + \rho e_t \, . \\
&= \left[\lambda^{-1}(\lambda - L) C_1(L) + \rho [1;\ 0] L C_2(L)\right] w_{t+1} \\
&= \ -(1-\lambda)\ [1;\ 0]\, D_2(\lambda) w_{t+1} \, .
\end{aligned} \tag{2.10}
$$

The implications of (2.10) are investigated further in Chapter 5.

Example 4: Martingale Model with Temporal Aggregation

As noted in Hansen and Singleton (1983, 1990), Grossman, Melino and Shiller (1987) and Hall (1988), several intertemporal asset pricing models have the implication that some linear combination of a vector of variables observed by an econometrician is a martingale difference sequence. Denote this linear combination by $\gamma \cdot y_t$ where $\gamma' = [\gamma_1\ \gamma_2']$ is a q-dimensional vector and γ_1 is different from zero. Hence

$$
\mathrm{P}\,(\gamma \cdot y_t | \Omega_{t-1}) = 0 \, . \tag{2.11}
$$

This model becomes a special case of the general model presented in section one by letting $A(z)$ be a scalar given by the first entry of γ, $B(z)$ be a $(q-1)$-dimensional row vector given by the remaining $(q-1)$-dimensional subvector of γ', and ℓ be one.

Instead of observing y_t directly, suppose that an econometrician observes a temporally-aggregated version of y_t. As in Grossman, Melino and Shiller (1987), Hall (1988), and Hansen and Singleton (1990), suppose that

$$
\gamma \cdot y_t = \int_0^1 \alpha(\tau) \cdot dW(t - \tau) \tag{2.12}
$$

where W is a continuous-time, vector martingale with stationary increments, and α is vector-valued, continuous function on [0, 1]. Relation (2.12) clearly implies that the discrete time process $\gamma \cdot y_t$ is a martingale difference sequence. Let y_t^a denote the temporally aggregated version of y_t. Then

$$\gamma \cdot y_t^a = \int_0^1 \left[\int_0^1 \alpha(\tau) dW(t - \tau - s) \right] ds \,. \tag{2.13}$$

As emphasized by Hall (1988), it follows from (2.13) that

$$\mathrm{P}\,(\gamma \cdot y_t^a \mid \Omega_{t-2}) = 0 \,. \tag{2.14}$$

In other words, temporal aggregation has the effect of shifting the information set back one additional time period so that ℓ is set to two. Thus $C(z)$ for the model with temporal aggregation is given by

$$\begin{aligned} C_1(z) &= (\gamma_1)^{-1} \left[D_1(z) - \gamma_2' \, D_2\,(z) \right] \\ C_2(z) &= D_2(z) \end{aligned} \tag{2.15}$$

where $D_1(z)$ is a first-order polynomial.

Example 5: Dynamic Demand Function for a Factor of Production

Sargent (1978) and Kennan (1979) have studied linear demand functions for factors of production that are derived from optimizing a quadratic objective function subject to linear constraints. We focus on Sargent's version, though a reinterpretation of the variables will yield Kennan's model as well. Assuming that there are no shocks to the technology and a single factor of production, the demand function turns out to be

$$n_t = \delta n_{t-1} - \theta \, \mathrm{P} \left[\sum_{j=0}^{\infty} (\beta\delta)^\tau p_{t+j} \mid \Omega_t \right] \tag{2.16}$$

where $0 < \beta < 1$, $0 < \delta < 1$, $\theta > 0$, n_t is the amount of the factor demanded at date t and p_t is the rental rate of the factor at date t. This model is of the form described in section one with $y_{1t} = n_t$, the first entry of y_{2t} equal to p_t, $A(z) = (1 - \delta z)$, $B(z) = 1/(1 - \beta\delta z)\,[\theta;\ 0]$ and $\ell = 0$. Then $C(z)$ satisfies:

$$\begin{aligned} C_1(z) &= [-\theta;\ 0]\ [zD_2(z) - \beta\delta D_2(\beta\delta)]/[(1 - \delta z)\,(z - \beta\delta)]; \\ C_2(z) &= D_2(z) \,. \end{aligned} \tag{2.17}$$

Multiple factor versions of this example can be constructed easily along the lines of Hansen and Sargent (1981a) and Kollintzas (1985).

The list of examples could be extended to incorporate linear rational expectations models which have been used to study a wide variety of macroeconomic and microeconomic phenomena. The preceding examples are sufficient to illustrate the variety of *exact* linear rational expectations models.

3. Identification

As in section one, we study models with vector moving-average representations

$$y_t = C(L)w_t \tag{3.1}$$

where y is a q-dimensional covariance stationary process and $C(z)$ satisfies (1.19) for some $H(z)$ satisfying *R1*. In other words, $C(z)$ satisfies the cross-equation restrictions implied by the underlying economic model. Since the coefficients of the power series expansion of $C(z)$ are square summable, $C(z)$ can be extended from the interior of the unit disk to the unit circle by letting

$$C[\exp(-i\theta)] = \sum_{j=0}^{\infty} c_j \exp(-i\theta j) \tag{3.2}$$

be defined almost everywhere on $[-\pi, \pi]$ as a matrix of mean-square limits.[3] The spectral density of y is given by

$$S(\theta) \equiv C[\exp(-i\theta)]C[\exp(i\theta)]' \tag{3.3}$$

where the prime denotes transposition (but not complex conjugation). The spectral density generates the autocovariances of y via the formula:

$$E(y_t y'_{t-\tau}) = (1/2\pi) \int_{-\pi}^{\pi} \exp(i\theta\tau)S(\theta)d\theta\ . \tag{3.4}$$

Taken together, formulas (3.3) and (3.4) give the autocovariances of the process y as a function of $C(z)$ used in representing y as a moving average of a vector white noise.

Without constraining C, it is known that there are multiple choices of C that imply the same spectral density and hence the same sequence of autocovariances. Loosely put, there is an equivalence class of matrix functions C which can be generated one from another by post-multiplying an original C by functional counterparts to orthogonal matrices. It is not possible to distinguish members of this equivalence class on the basis of the implied autocovariances of the time series. In other words, there is an identification problem in representing y as a one-sided moving-average of a vector white noise of disturbances.

In many circumstances, especially in problems involving prediction, this identification problem is resolved in part by constructing a moving-average representation in terms of a process v that is fundamental for y:

$$y_t = F(L)v_t \,. \tag{3.5}$$

We refer to (3.5) as a *fundamental moving-average representation*. Given one fundamental moving-average representation characterized by a function $F(z)$, all other fundamental representations are obtained by post-multiplying F by orthogonal matrices U, i.e. by matrices of real numbers that satisfy $UU' = I$.

Once we relax the restriction that the moving-average representation be fundamental, there is a much larger class of observationally equivalent moving-average representations. Again these moving-average representations are obtained by post-multiplying $F(z)$ by $U(z)$ except that now $U(z)$ can be an orthogonal matrix function, an object that we now define.

Definition: $U(z)$ is said to be a q-dimensional *orthogonal matrix function* if $U(z)$ is a $(q \times q)$ matrix function of a complex variable with a power series expansion $U(z) = \sum_{j=0}^{\infty} u_j\, z^j$ satisfying

(i) the entries of u_j are real numbers for $j \in J^+$;
(ii) $\sum_{j=0}^{\infty} \text{trace}\,(u_j u_j') < \infty$;
(iii) $U[\exp(-i\theta)]\, U[\exp(i\theta)]' = I$ almost everywhere.

Given a matrix function $F(z)$ determining a fundamental moving-average representation and a q-dimensional orthogonal matrix function $U(z)$, we can construct other moving-average representations via the formula:

$$C(z) = F(z)U(z) \tag{3.6}$$

[e.g. see Rozanov (1967), page 62]. In light of (i), the coefficient matrices of the power series expansion of C have entries that are real numbers. In light of (ii) and (iii), the boundary values of C on the unit circle satisfy:

$$\begin{aligned} &C[\exp(-i\theta)]\, C[\exp(+i\theta)]' = \\ &\quad F[\exp(-i\theta)]\, U[\exp(-i\theta)]\, U[\exp(i\theta)]'\, F[\exp(i\theta)]' \\ &\quad = F[\exp(-i\theta)]\, F[\exp(i\theta)]' \,. \end{aligned} \tag{3.7}$$

Consequently, it follows from (3.3) and (3.4) that the spectral density function and autocovariance sequence implied by $C(z)$ is identical to that implied by $F(z)$. As a result, for a given fundamental representation (3.5), we can construct alternative observationally equivalent moving-average representations of the form (3.1) by selecting a q-dimensional orthogonal matrix function $U(z)$, forming $C(z)$ as in (3.6), and forming w_t via

$$w_t = U(L^{-1})' v_t \tag{3.8}$$

where

$$U(L^{-1})' = \sum_{j=0}^{\infty} u_j' L^{-j} \ . \tag{3.9}$$

Note that (iii) implies that $U(L^{-1})' = U(L)^{-1}$. From (3.8) it follows that $v_t = U(L)w_t$, so that v_t depends only on current and past values of w_t. Therefore, the linear space Λ_t generated by current and past values of w_t is at least as large as the linear space Σ_t generated by current and past values of v_t. Whenever u_j is different from zero for some $j \geq 1$, Λ_t is strictly larger than Σ_t.

Consider the class of $C(z)$'s constructed via (3.6) and indexed by alternative q-dimensional orthogonal functions $U(z)$. Suppose that at least one member of this class, say $\hat{C}(z)$, satisfies (1.19) for some $\hat{H}(z)$ satisfying *R1*. Not all of the remaining members of this class of $C(z)$'s will necessarily satisfy the cross-equation restrictions. Hence the cross-equation restrictions go at least part of the way in reducing the class of observationally equivalent moving-average representations. The remainder of this section studies the extent of this reduction.

Put somewhat differently, we are interested in characterizing the class of observationally equivalent $D(z)$'s. Formula (1.14) describes the mapping from *admissible* $D(z)$'s, i.e. $D(z)$'s for which $D_1(z)$ is a polynomial with degree $\ell - 1$, to the class of $C(z)$'s that satisfy the restrictions. Formulas (3.3) and (3.4) delineate the mapping from the class of $C(z)$'s used in moving-average representations for y to the family of spectral density functions and autocovariance sequences for y. Taken together, these formulas give a mapping from the space of admissible $D(z)$'s to the space of admissible autocovariances for y. The identification question we analyze pertains to the inverse of this mapping. What is the class of admissible $D(z)$'s associated with a given admissible autocovariance sequence?

Since we focus only on admissible autocovariance sequences, there exists at least one admissible $D(z)$ that generates this sequence, say $\hat{D}(z)$. Corresponding to this $\hat{D}(z)$ is a $\hat{C}(z)$ and an $\hat{H}(z)$ given by formula (1.20). Consider any other $C(z) = F(z)U(z)$ for some q-dimensional orthogonal matrix function $U(z)$. The matrix functions $C(z)$ and $\hat{C}(z)$ are related via:

$$C(z) = \hat{C}(z)\,\hat{U}(z^{-1})'\,U(z)\,. \tag{3.10}$$

Since $\hat{C}(z)$ and $\hat{H}(z)$ satisfy equation (1.19),

$$[A(z);\ B(z^{-1})]\,\hat{C}(z) = z^{\ell}\,\hat{H}(z^{-1})\,. \tag{3.11}$$

Post-multiply both sides of (3.11) by $\hat{U}(z^{-1})'\,U(z)$ to obtain

$$[A(z);\ B(z^{-1})]\,C(z) = z^{\ell}\hat{H}(z^{-1})\,\hat{U}(z^{-1})'\,U(z)\,. \tag{3.12}$$

Therefore, $C(z)$ satisfies the restrictions if and only if:

Restriction R2: $\hat{H}(z)\,\hat{U}(z)'\,U(z^{-1})$ is analytic on the domain $\{z : |z| < 1\}$.

We have established:

Lemma 1: Suppose that $C(z) = F(z)U(z)$ for some q-dimensional orthogonal matrix function $U(z)$. Then $C(z)$ satisfies (1.14) for some admissible $D(z)$ if, and only if $U(z)$ satisfies *R2*.

Since $\hat{H}(z)$ is analytic inside the unit circle of the complex plane, one convenient sufficient condition for *R2* is:

Restriction R3: $\hat{U}(z)'\,U(z^{-1})$ is analytic on $\{z : |z| < 1\}$.

Restriction *R3* is always satisfied when U is a constant orthogonal matrix independent of z. Hence the restrictions are always satisfied for fundamental moving-average representations. More generally, note that

$$\begin{aligned} y_t &= \hat{C}(L)\hat{w}_t \\ &= C(L)\,U(L^{-1})'\,\hat{U}(L)\hat{w}_t \\ &= C(L)w_t\,. \end{aligned} \tag{3.13}$$

Consequently, the w process must satisfy:

$$w_t = U(L^{-1})'\,\hat{U}(L)\,\hat{w}_t\,. \tag{3.14}$$

When restriction *R3* is satisfied, the entries of w_t are in the closed linear space $\hat{\Lambda}_t$ generated by current and past values of $\hat{w}_t$. Hence the closed linear space Λ_t generated by current and past values of w_t is no larger than $\hat{\Lambda}_t$. It follows from Lemma 1 that the restrictions are preserved for any moving-average representation associated with an information reduction. Alternatively, this special case of Lemma 1 could be proved by assuming that (1.8) is satisfied for the information set $\hat{\Lambda}_{t-\ell}$ and applying the Law of Iterated Projections (see Chapter 2, Lemma 3.5) to show that (1.8) is also satisfied for the smaller information set $\Lambda_{t-\ell}$.

Restriction *R2* turns out to be weaker than restriction *R3*. Thus Lemma 1 covers cases other than those associated with an information reduction. To see this, consider the following example.

Example 1: Let λ be any real number such that $|\lambda| < 1$. Form the spectral decomposition of the symmetric positive semidefinite matrix

$$\hat{H}(\lambda)'\,\hat{H}(\lambda) = U_1 V U_1' \tag{3.15}$$

where $U_1 U_1' = U_1' U_1 = I$ and V is a diagonal matrix with strictly positive real numbers in the diagonal entries of its first block and zeros in the second block. Then the second column block of $\hat{H}(\lambda)U_1$ contains only zeros. As in Rozanov (1967, page 47), we construct an orthogonal matrix function $U_2(z)$ of the form

$$U_2(z) = \begin{bmatrix} I & 0 \\ 0 & \beta(z)I \end{bmatrix} \tag{3.16}$$

where $\beta(z)$ is the *Blaschke* factor

$$\beta(z) \equiv (z-\lambda)/(1-\lambda z)\ . \tag{3.17}$$

Notice that

$$\begin{aligned} \beta(z)\beta(z^{-1}) &= (z-\lambda)\ (z^{-1}-\lambda)/[(1-\lambda z)\ (1-\lambda z^{-1}]\ . \\ &= 1\ . \end{aligned} \tag{3.18}$$

Consequently, $U_2(z)$ is a q-dimensional orthogonal matrix function. Let

$$U(z) = \hat{U}(z)\,U_1 U_2(z)\ . \tag{3.19}$$

Then $U(z)$ also is a q-dimensional orthogonal matrix function. Note that

$$\hat{H}(z)\hat{U}(z)'U(z^{-1}) = \hat{H}(z)U_1U_2(z^{-1})\ . \tag{3.20}$$

Although $U_2(z^{-1})$ has a pole at $z = \lambda$, $\hat{H}(z)U_1U_2(z^{-1})$ is analytic on $\{z : |z| < 1\}$ because the singularity at $z = \lambda$ is removable by construction. This follows from the fact that $\hat{H}(\lambda)U_1$ has all zeros in its second column block whereas the pole of $U_2(z^{-1})$ at $z = \lambda$ occurs in its second row block. Consequently, $U(z^{-1})$ given by (3.19) satisfies *R2*. Furthermore,

$$\hat{U}(z)'U(z^{-1}) = U_1U_2(z^{-1}) , \tag{3.21}$$

which has a pole at $z = \lambda$ that is not removable. Therefore, $U(z)$ fails to satisfy *R3*.

Following the construction in Example 1, a rich family of $U(z)$'s can be generated, each member of which satisfies *R2* but fails to satisfy *R3*. For instance, any real value of λ for which $|\lambda| < 1$, can be used. In addition, given any finite number of $U(z^{-1})$'s that satisfy *R2* but fail to satisfy *R3*, the product of these $U(z^{-1})$'s possesses this same property. Finally, a construction similar to that given in Example 1 will work for complex values of λ with some minor modifications. First, the matrix $\hat{H}(\lambda)'$ on the left side of (3.15) should be replaced by its conjugate in constructing a (conjugate) symmetric matrix that is positive semidefinite. Second, the coefficient matrices of the power series expansion of the resulting $U(z)$ may not have all real entries. However, by repeating the construction using the complex conjugate of λ in the second stage and post multiplying by an appropriately chosen unitary matrix, this complication can be avoided. In summary, there exists a rich collection of q-dimensional orthogonal matrix functions that satisfy *R2* but not *R3*.

In practice, one typically adopts a finite-dimensional parameterization of $\hat{D}_2(z)$. It is of interest to see what impact this specification has on the analysis. Suppose that $\hat{D}_2(z)$ is given by the ratio of polynomials:

$$\hat{D}_2(z) = \hat{D}_n(z)/\hat{D}_d(z) \tag{3.22}$$

where $\hat{D}_n(z)$ and $\hat{D}_d(z)$ are finite-order polynomials and the zeros of $\hat{D}_n(z)$ are all outside the unit circle. Let N denote the order of $\hat{D}_n(z)$. Suppose we admit only $C(z)$'s constructed from $D_2(z)$'s of the form:

$$D_2(z) = D_n(z)/\hat{D}_d(z) \tag{3.23}$$

where the order of $D_n(z)$ does not exceed N. Our candidate for $D_n(z)$ is given by

$$D_n(z) = \hat{D}_n(z)\hat{U}(z^{-1})'U(z) . \tag{3.24}$$

By assumption, $D_2(z)$ and hence the right side of (3.24) is analytic inside the unit circle of the complex plane. However, the right side of (3.24) will not necessarily be a polynomial. Hence the restriction that $D_n(z)$ be a finite order polynomial with the same degree as $\hat{D}_n(z)$ limits further the class of admissible $C(z)$'s.

Consider first the case in which the more stringent restriction *R3* is satisfied. In this case the right side of (3.24) is analytic everywhere in the complex plane. Furthermore, since $\hat{U}(z^{-1})'U(z)$ has a one-sided Laurent series expansion in terms of negative powers of z, it can be extended to be analytic at the point ∞. Therefore

$$\lim_{z \to \infty} z^{-N+1} \hat{D}_n(z) \hat{U}(z^{-1})' U(z) = 0 . \tag{3.25}$$

It follows that $\hat{D}_n(z)\hat{U}(z^{-1})' U(z)$ has a finite order pole at ∞ and that the order of this pole does not exceed N. Therefore, $\hat{D}_n(z)\hat{U}(z^{-1})' U(z)$ is a finite order polynomial and its order does not exceed N. Hence requiring that $D_n(z)$ be an Nth order polynomial does not alter the analysis of *R3*.

We consider next the more general restriction *R2*, and ask whether *R2* can be satisfied even though *R3* is violated.

Example 2: We show how to modify the construction in Example 1 to accommodate the order restriction on $D_n(z)$. Since $A(z)$, $B(z)$ and $D(z)$ are matrices of rational functions, so is $\hat{H}(z)$. In Example 1 we exploited the fact that the matrix $\hat{H}(\lambda)' \hat{H}(\lambda)$ is always singular. However, in this example we must confine our attention to only a finite number of values of λ. In particular, let

$$\det \begin{bmatrix} \hat{H}(z^{-1}) \\ \hat{D}_n(z) \end{bmatrix} = \hat{\delta}_n(z)/\hat{\delta}_d(z) \tag{3.26}$$

where $\hat{\delta}_n(z)$ and $\hat{\delta}_d(z)$ do not have any zeros in common. Suppose that there exists a real number λ such that $|\lambda| < 1$ and λ^{-1} is a zero of $\hat{\delta}_n(z)$ but not a zero of $\hat{D}_d(z)$ [i.e., λ is not a pole of $\hat{D}_2(z)$]. Such a λ does not always exist, although it often does.

As in Example 1, we use λ to construct $U(z)$. In this case we begin by forming the spectral decomposition of the symmetric, positive semi-definite matrix:

$$\hat{H}(\lambda)' \hat{H}(\lambda) + \hat{D}_n(\lambda^{-1})' \hat{D}_n(\lambda^{-1}) = U_1 V U_1' . \tag{3.27}$$

At least one entry of V, the bottom right one, must be zero because λ^{-1} is a zero of $\hat{\delta}_n(z)$. Note that by construction, the second column blocks of $\hat{H}(\lambda)U_1$ and $\hat{D}_n(\lambda^{-1})U_1$ are zero. Use the same orthogonal matrix functions $U_2(z)$ and $U(z)$ that were used in Example 1:

$$\hat{D}_n(z)\hat{U}(z^{-1})'U(z) = \hat{D}_n(z)U_1U_2(z)\,. \tag{3.28}$$

As in Example 1, $H(z)$ is analytic inside the unit circle. However, in this case we also preserve the requirement that $D_n(z)$ be a polynomial with an order no greater than N because the second column block of $\hat{D}_n(\lambda^{-1})U_1$ is zero and the first order pole of $U_2(z)$ occurs in the second row block. Consequently, the singularity of $\hat{D}_n(z)U_1U_2(z)$ at λ^{-1} is removable.

Like Example 1, Example 2 assumes that λ is real. However, $\hat{\delta}_n(z)$ may have complex zeros that exceed one in modulus and are not simultaneously zeros of $\hat{D}_d(z)$. Such zeros come in complex conjugate pairs. These conjugate pairs of zeros can also be used in an analogous (but slightly more complicated) fashion to construct orthogonal matrix functions that respect the order restriction on $D_n(z)$. The order restriction does limit us to a finite number of choices of λ because $\hat{\delta}_n(z)$ has a finite number of zeros. Finite products of such orthogonal matrix functions can be formed so long as any given λ is not used more times than its multiplicity as a zero of $\hat{\delta}_n(z)$.

Example 3: We show by illustration that it is possible for $\hat{\delta}_n(z)$ to have a zero outside the unit circle of the complex plane that is not simultaneously a zero of $\hat{D}_d(z)$. Set $A(z) = 1$, $B(z^{-1}) = z^{-1}$, $\ell = 0$, $\hat{D}_n(z) = [1 + 2z + z^2,\ 1]$ and $\hat{D}_d(z) = 1$. Hence $n = 2$, $\hat{D}_1(z) = 0$, and $\hat{H}(z^{-1}) = [1 \quad 1]$. Note that

$$\det\begin{bmatrix}\hat{H}(z^{-1})\\ D_n(z)\end{bmatrix} = -2z - z^2 = \hat{\delta}_n(z)\,. \tag{3.29}$$

The polynomial $\hat{\delta}_n(z)$ has one zero outside the unit circle, namely $z = -2$. Therefore, $\lambda = -1/2$ can be used in the construction described in Example 2.

In summary, in this section we have investigated the extent to which the cross-equation restrictions implied by members of the class of exact rational expectations models can be used to narrow the family of observationally equivalent moving-average representations. We have found that the function $D_2(z)$ that can be used to index alternative solutions is not identified. Even when entries of $D_2(z)$ are restricted to be ratios of polynomials with prespecified orders, $D_2(z)$ can fail to be identified.

The standard approach in reduced-form time series analysis is to focus exclusively on fundamental moving-average representations. As we have indicated, fundamental representations will satisfy the restrictions implied by the model. Requiring that $C(z)$ be fundamental is equivalent to the restriction that

$$C(z) = F(z)U \tag{3.30}$$

for some orthogonal matrix U that is independent of z. However, it may be computationally tedious to restrict the family of $D(z)$'s so that resulting $C(z)$'s satisfy (3.30). Furthermore, this extra restriction is *ad hoc*, and the underlying information structure faced by economic agents is still left unidentified.[4]

While these observations are discouraging from the standpoint of identifying $D_2(z)$, they do not overturn the validity of likelihood-based inferences. The restricted and unrestricted likelihood functions will typically have multiple peaks, but the peaks of interest will all have the same value.

The analysis in this section has taken $A(z)$ and $B(z^{-1})$ as given and has focused on the identification of $D(z)$ and hence $C(z)$. More generally, $A(z)$ and $B(z^{-1})$ may only be known up to a finite-dimensional parameter vector. It is often possible to identify the parameters governing $A(z)$ and $B(z^{-1})$ even though identification of $D(z)$ is problematic. When $A(z)$ and $B(z)$ are not identified, the family of observationally equivalent $D(z)$'s is likely to be expanded.

4. Restrictions Implied for First Differences

In the analysis considered thus far, we have assumed that the y process is covariance stationary. Although we could view this assumption as being appropriate for deviations about a linear time trend, an alternative strategy is to assume that the first difference of y_2 is covariance stationary where the process $y' \equiv [y_1' \; y_2']$ is partitioned in a manner compatible with $C(z)$. We maintain the model restrictions (1.3) which for convenience are written below:

$$P\Big\{[A(L) \;;\; B(L^{-1})]\, y_t | \Omega_{t-\ell}\Big\} = 0 \,. \tag{4.1}$$

Recall that $B(z) = \sum_{j=0}^{\infty} b_j z^j$. Let

$$(4.2)\qquad \begin{aligned} b_j^* &\equiv \sum_{k=j}^{\infty} b_j \quad \text{for } j = 0, 1, \ldots\,, \\ y_{2t}^* &\equiv y_{2t} - y_{2,t-1} \end{aligned}$$

and

$$B^*(z) \equiv \sum_{j=1}^{\infty} b_j^*(z)\,.$$

Now $b_j = b_j^* - b_{j+1}^*$, implying that[5]

$$(4.3)\qquad B(L^{-1})y_{2t} = B^*(L^{-1})y_{2t}^* + b_0^*\, y_{2t}\,.$$

Substituting (4.3) into (4.1), we obtain

$$(4.4)\qquad P\Big[A(L)y_{1t} + b_0^*\, y_{2t} + B^*(L^{-1})y_{2t}^* | \Omega_{t-\ell}\Big] = 0$$

Define

$$(4.5)\qquad y_{1t}^* \equiv A(L)y_{1t} + b_0^*\, y_{2t}\,.$$

Then we can write the first-difference model as

$$(4.6)\qquad P\Big\{[I\ ;\ B^*(L^{-1})]y_t^* | \Omega_{t-\ell}\Big\} = 0$$

where $y^{*\prime} \equiv [y_1^{*\prime}\ ,\ y_2^{*\prime}]$ is assumed to be covariance stationary. This is just a special case of the general model presented in Section 1.

The first-difference model derived here is usefully compared to that employed by Sargent (1979). In particular, Sargent first differenced (4.1) and projected both sides onto $\Omega_{t-\ell-1}$ to obtain

$$(4.7)\qquad P\Big\{[A(L)\ ;\ B(L^{-1})]\ (y_t - y_{t-1}) | \Omega_{t-\ell-1}\Big\} = 0\,.$$

Although restrictions (4.7) can be tested using procedures discussed in this paper, some implications of (4.1) are lost by projecting onto $\Omega_{t-\ell-1}$ rather than $\Omega_{t-\ell}$. On the other hand, (4.6) involves a projection onto $\Omega_{t-\ell}$ rather than $\Omega_{t-\ell-1}$ and imposes more restrictions than (4.7). Therefore, it is quite possible that the procedures proposed in this section can detect empirical contradictions of the hypothesis (4.1) that Sargent's procedure could not.

5. Likelihood Estimation and Inference

In this section we describe briefly how to conduct estimation and inference using the method of maximum likelihood with a Gaussian likelihood function. First we illustrate how to impose the restrictions implied by the model on the moving-average coefficients. Then we describe two alternative approaches to evaluating the likelihood function.

Imposing the Restrictions

For pedagogical purposes, we focus on the following special case of the model described in section one. Suppose $A(z) = I$ and $B(z) = b_0 + b_1 z + \ldots + b_k z^k$ and $\ell = 0$. Furthermore, we consider rational parameterizations of $D_2(z)$ of the form described in Section 3:

$$D_2(z) = D_n(z)/D_d(z) \tag{5.1}$$

where $D_n(z) = d_{n0} + d_{n1}z + \ldots + d_{nk}z^k$ and $D_d(z) = d_{d0} + d_{d1}z + \ldots + d_{dk}z^k$. The restrictions that the polynomials $B(z)$, $D_n(z)$ and $D_d(z)$ all have the same order is made only for notational convenience. Additional zero restrictions can be imposed on these polynomials in a straightforward fashion. To guarantee that $D_d(z)$ has a power series expansion with square summable coefficients, we restrict the zeros of $D_d(z)$ to be outside the unit circle of the complex plane. A convenient way to accomplish this is to use a parameterization suggested by Monahan (1984).

For the special case of the model considered here, (1.18) simplifies to

$$\begin{aligned} C_1(z) &= -B(z^{-1})\, D_2(z) + G(z^{-1}) \\ C_2(z) &= D_2(z) \end{aligned} \tag{5.2}$$

where $G(z)$ satisfies *R1*. Substituting (5.1) into (5.2) yields

$$C_1(z)D_d(z) = -B(z^{-1})\, D_n(z) + D_d(z)G(z^{-1})\,. \tag{5.3}$$

There are two important implications of (5.3). First, $C_1(z)D_d(z)$ is a finite-order polynomial with maximum order k. This follows from the fact that the largest positive power of z that occurs on the right side of (5.3) is k. Hence

$$C_1(z) = C_n(z)/D_d(z) \tag{5.4}$$

for some k^{th}-order polynomial $C_n(z)$. Second, $G(z)$ is a finite order polynomial with maximum order k. This follows from the fact that the largest (in absolute value) negative power of z that occurs in $B(z^{-1})$ is $-k$.

Substituting (5.4) into (5.3) gives

$$C_n(z) = -B(z^{-1})D_n(z) + D_d(z)G(z^{-1}) \tag{5.5}$$

which can be viewed as a system of linear equations in the coefficients of $C_n(z)$ and $G(z)$. Furthermore, there is a recursive structure to these equations as evident by the fact that the equations determining the coefficients of $G(z)$ do not involve the coefficients of $C_n(z)$. Written in matrix notation, the first set of equations is

$$\begin{aligned}\begin{bmatrix}0\\0\\\vdots\\0\end{bmatrix} &= -\begin{bmatrix}b_k & 0 & \dots & 0\\ b_{k-1} & b_k & \dots & 0\\ \vdots & \vdots & \ddots & \vdots\\ b_1 & b_2 & \dots & b_k\end{bmatrix}\begin{bmatrix}d_{n0}\\d_{n1}\\\vdots\\d_{nk-1}\end{bmatrix}\\ &+\left\{\begin{bmatrix}d_{d0} & 0 & \dots & 0\\ d_{d1} & d_{d0} & \dots & 0\\ \vdots & \vdots & \ddots & \vdots\\ d_{d,k-1} & d_{d,k-2} & \dots & d_{d0}\end{bmatrix}\otimes I\right\}\begin{bmatrix}g_k\\g_{k-1}\\\vdots\\g_1\end{bmatrix}\end{aligned} \tag{5.6}$$

where $G(z) = g_1z + g_2z^2 + \ldots + g_kz^k$. Since the matrix in $\{\cdot\}$ in (5.6) is nonsingular, this system of equations can be solved for the coefficients of G. Given these coefficients, one can then compute the coefficients of $C_n(z)$ by

$$\begin{aligned}\begin{bmatrix}c_{n0}\\c_{n1}\\\vdots\\c_{nk}\end{bmatrix} &= -\begin{bmatrix}b_0 & b_1 & \dots & b_k\\ 0 & b_0 & \dots & b_{k-1}\\ \vdots & \vdots & \ddots & \vdots\\ 0 & 0 & \dots & b_0\end{bmatrix}\begin{bmatrix}d_{n0}\\d_{n1}\\\vdots\\d_{nk}\end{bmatrix}\\ &+\left\{\begin{bmatrix}d_{d1} & d_{d2} & \dots & d_{dk}\\ d_{d2} & d_{d3} & \dots & 0\\ \vdots & \vdots & \ddots & \vdots\\ 0 & 0 & \dots & 0\end{bmatrix}\otimes I\right\}\begin{bmatrix}g_1\\g_2\\\vdots\\g_k\end{bmatrix}\end{aligned} \tag{5.7}$$

Therefore, the coefficients of $C_n(z)$ can be computed using some simple matrix manipulations.[6]

Likelihood Evaluation

A Gaussian likelihood function can be expressed as a function of the mean vector and covariance matrix of the data, (say $y_1, y_2, \ldots, y_T$), and of the unknown parameters. In this paper we have abstracted from restrictions on the mean vector, and have instead investigated the covariance restrictions. These covariance restrictions can be viewed equivalently as restrictions on the spectral density of y. In sections one, three and the first part of this section, we described the mapping from $D(z)$ to the spectral density matrix of y. This representation of the restriction makes it convenient to follow the suggestions of Hannan (1970), Robinson (1977), Dunsmuir and Hannan (1976), and Dunsmiur (1978) by using a frequency domain approximation to the likelihood function.

This approximation is formed as follows. First, form the finite Fourier transform of the y sequence:

$$Y(\theta_j) = \sum_{t=1}^{T} y_t \, e^{-i\theta_j t} \tag{5.8}$$

where $\omega_j = \frac{2\pi j}{T}$, $j = 1, 2, \ldots, T-1$. We omit frequency zero from consideration since sample means are subtracted from our time series. The periodogram is defined as

$$I(\theta_j) = \frac{1}{T} Y(\theta_j) Y(-\theta_j)' \, . \tag{5.9}$$

Then the log likelihood of the sample $\{y_t : t = 1, \ldots, T\}$ is approximated by

$$\begin{aligned} L = \frac{nT}{2} \log 2\pi - \frac{1}{2} \sum_{j=1}^{T-1} \log \det S(\theta_j) \\ - \frac{1}{2} \sum_{j=1}^{T-1} \text{trace } S(\theta_j)^{-1} I(\theta_j) \end{aligned} \tag{5.10}$$

where $S(\theta_j)$ is the theoretical spectral density defined in (3.3) and (2.3). Note that $S(\theta_j)$ can be expressed in terms of the free parameters of $D(z)$ using formulas (1.14) and (3.3). To emphasize the dependence on $C(z)$ and hence on $D(z)$, it is worthwhile substituting (3.3) into (5.10) to obtain

$$
(5.11)\quad \begin{aligned} L = &- \frac{nT}{2} \log 2\pi - \frac{1}{2} \sum_{j=1}^{T-1} \log \det [C(e^{-i\theta_j}) C(e^{i\theta_j})'] \\ &- \frac{1}{2} \sum_{j=1}^{T-1} \text{trace } [C(e^{-i\theta_j}) C(e^{i\theta_j})'] I(\theta_j) . \end{aligned}
$$

In computing (5.11), it is useful to exploit the fact that

$$
\log \det[C(e^{-i\theta})C(e^{i\theta})'] = \log \det\{C[e^{-i(2\pi-\theta)}]C[e^{i(2\pi-\theta)}]'\} ,
$$

and

$$
(5.12)\quad \begin{aligned} &\text{trace } [C(e^{-i\theta}) C(e^{i\theta})']^{-1} I(\theta) \\ &= \text{trace } \{C[e^{-i(2\pi-\theta)}] C[e^{i(2\pi-\theta)}]'\}^{-1} I(2\pi - \theta) . \end{aligned}
$$

These formulas permit (5.10) to be rewritten in terms of sums over only $T/2$ frequencies. The free parameters of $D(z)$ are estimated by maximizing (5.11) subject to the restrictions (1.14).

Alternatively, by restricting $D(z)$ appropriately, one can often avoid making approximations to the likelihood function and instead use filtering methods to evaluate it. For instance, consider the model and the parameterization of $D(z)$ given in the first subsection. In this case, $C(z)$ can be represented as

$$
(5.13)\quad \begin{aligned} C_1(z) &= C_n(z)/D_d(z) \\ C_2(z) &= D_n(z)/D_d(z) \end{aligned}
$$

where $C_n(z)$, $D_n(z)$ and $D_d(z)$ are finite-order polynomials. In this case, the y process is a vector autoregressive moving-average (ARMA) process. It is known that such a process has a state-space representation [e.g. see Anderson and Moore (1979), page 236].

Recall that the joint density of $y_1, y_2, \ldots, y_T$ can be expressed as a product of conditional densities, say the product of the density of y_T conditioned on $y_1, y_2, \ldots, y_{T-1}$, the density of y_{T-1} conditioned on $y_1, y_2, \ldots, y_{T-2}, \ldots$, and the marginal density of y_1. Each of these densities is Gaussian and hence can be constructed given knowledge of the conditional means and covariance matrices. Given the state-space representation, one can calculate recursively the required conditional expectations and conditional covariance matrices using one of several possible filtering algorithms described in Anderson and Moore (1979).

The initial covariance matrix required for these algorithms can be computed using a doubling algorithm [see Anderson and Moore (1979), page 67].[7]

The analysis in this section took $A(z)$ and $B(z^{-1})$ as given and focused on the identification of $D(z)$ and hence $C(z)$. If $A(z)$ and $B(z^{-1})$ are only known up to a finite-dimensional parameter vector, it will only be more difficult to identify of $D(z)$. However, even if $D(z)$ is not fully identified, it is often possible to identify the parameters governing $A(z)$ and $B(z^{-1})$.

6. Inexact Models with Hidden Variable Interpretations of Disturbances

So far, this paper has been confined to analyzing exact linear rational expectations models. In this section, we briefly indicate which aspects of the analysis readily carry over to more general linear rational expectations models and which aspects require modification. It turns out that the methods of representing the cross-equation restrictions and estimating the parameters both carry over with minimal modification. However, the treatment of identification must be modified substantially.

Following Sargent (1978) and Hansen and Sargent (1980a), we assume that a subvector of y_{2t} is not observed by the econometrician. For this reason, we partition y_{2t} as $y_{2t}' = [y_{2t}^{o\prime}, y_{2t}^{u\prime}]$ where the process y_2^o is observed by the econometrician while the process y_2^u is not. We partition $C(z)$ accordingly, so that the moving-average representation for y is:

$$\begin{bmatrix} y_{1t} \\ y_{2t}^o \\ y_{2t}^u \end{bmatrix} = \begin{bmatrix} C_1(L) \\ C_2^o(L) \\ C_3^u(L) \end{bmatrix} w_t \, . \tag{6.1}$$

Similarly, we partition $B(z) = [B^o(z);\ B^u(z)]$. Hence we write equation (1.3) as

$$A(L)y_{1t} + \mathrm{P}\left[B^o(L^{-1})y_{2t}^o|\Omega_t\right] = -\,\mathrm{P}\left[B^u(L^{-1})y_{2t}^u|\Omega_t\right] \tag{6.2}$$

where we have assumed that $\ell = 0$. We refer to such a model as an *inexact* rational expectations model because of the term $\mathrm{P}\left[B^u(L^{-1})y_{2t}^u|\Omega_t\right]$. The introduction of this term means that the relationship observed by the econometrician would not be exact even if economic agents could forecast perfectly.

From the vantage of model specification and solution, an inexact model as given here is identical with an exact model. However, from the vantage point of identification and testable restrictions, an inexact model can be quite different. For instance, suppose that $B^u(z)$ is nonsingular inside and on the unit circle of the complex plane, implying that there are r entries of y_2^u. Partition $D_2(z)$ as

$$D_2(z) = \begin{bmatrix} D_2^o(z) \\ D_2^u(z) \end{bmatrix} . \tag{6.3}$$

Then

$$\mathrm{P}\,[B^u(L^{-1})y_{2t}^u | \Lambda_t] = K^u(L)w_t \tag{6.4}$$

where

$$K^u(z) \equiv [B^u(z^{-1})D_2^u(z)]_+ \, . \tag{6.5}$$

Inverting relationship (6.5), we have that

$$D_2^u(z) = [B^u(z^{-1})^{-1}\, K^u(z)]_+ \, . \tag{6.6}$$

Therefore, without restricting $D_2^u(z)$, there are no restrictions on $K^u(z)$, and hence relation (6.2) imposes no restrictions on the process $[y_1', \, y_2^{o\prime}]'$ observed by the econometrician. Consequently, the empirical content of model (6.2) is tied to the *a priori* restrictions which are imposed on $D_2(z)$.

One way to restrict $D_2(z)$ is to require that it be block diagonal:

$$D_2(z) = \begin{bmatrix} D_2^o(z) \\ D_2^u(z) \end{bmatrix} = \begin{bmatrix} D_2^{oo}(z) & 0 \\ 0 & D_2^{uu}(z) \end{bmatrix} \tag{6.7}$$

where $D_2^{uu}(z)$ is a square matrix. In this case,

$$K^u(z) = [0 \quad [B^u(z^{-1})\, D_2^{uu}(z)]_+] \, , \tag{6.8}$$

and

(6.9)
$$\begin{bmatrix} C_1(z) \\ C_2^o(z) \end{bmatrix} = \begin{bmatrix} -A(z)^{-1}[B^o(z^{-1})D_2^{oo}(z)]_+ & -A(z)^{-1}[B^u(z^{-1})D_2^{uu}(z)]_+ \\ D_2^{oo}(z) & 0 \end{bmatrix}$$

The first column partition of the right side of (6.9) is

$$\begin{bmatrix} -A(z)^{-1}[B^o(z^{-1})D_2^{oo}(z)]_+ \\ D_2^{oo}(z) \end{bmatrix} \tag{6.10}$$

which is a square matrix and has the same structure as $C(z)$ given in (1.14). Consequently, the analysis in Section 3 applies to this component. Notice that the unobserved process y_2^u contaminates only the upper $(q \times q)$ portion of the spectral density of $[y_1', y_2^{o\prime}]'$.

Recall that the number of columns of $D_2(z)$ is equal to the number of entries of the stochastic process y. This restriction was imposed to ensure that y is stochastically nonsingular. Since $D_2^{uu}(z)$ is assumed to be a square matrix, $D_2^{oo}(z)$ has $q - r$ more columns than rows (where r is the dimension of y_1). A second way to restrict $D_2(z)$ is to treat y_2^u and y_2^o symmetrically by requiring also that $D_2^{oo}(z)$ be a square matrix. In this case, the y process is stochastically singular but the process $[y_1', y_2^{o\prime}]'$ observed by the econometrician is, in general, stochastically nonsingular. Now the matrix function in (6.10) is not square and the analysis in Section 3 will no longer apply.

It is of interest to investigate further this second restriction on $D_2(z)$. Given the structure of the matrix $\begin{bmatrix} C_1(z) \\ C_2^o(z) \end{bmatrix}$,

(6.11)
$$\det \begin{bmatrix} C_1(z) \\ C_2^o(z) \end{bmatrix} = - \det [D_2^{oo}(z)] \det \left\{ [-A(z)^{-1}] [B^u(z^{-1}) D_2^{uu}(z)]_+ \right\} .$$

Therefore, a zero of $\det \begin{bmatrix} C_1(z) \\ C_2^o(z) \end{bmatrix}$ must be either a zero of $\det [D_2^{oo}(z)]$ or a zero of $\det \{[-A(z)^{-1}[B^u(z^{-1}) D_2^{uu}(z)]_2\}$. As argued by Sargent (1978) and Hansen and Sargent (1980a), if $D_2^{oo}(z)$ is restricted to be nonsingular inside the unit circle of the complex plane, then y_1 should not cause y_2^0 in the sense of Granger (1969). This is one of the testable implications of the model.

Even when $D_2^{oo}(z)$ is nonsingular, moving-average representation (6.9) may fail to be fundamental because $\det \{-A(z)^{-1} [B^u(z^{-1}) D_2^{uu}(z)]_+\}$ may have zeros inside the unit circle of the complex plane. As emphasized by Hansen and Sargent (1980a), such zeros can exist even when $D_2^{uu}(z)$ is nonsingular at all points inside the unit circle. However, Hansen and Sargent (1980a) also note that one can construct a fundamental representation for the observable components of y that satisfies the restrictions. To see this, define

(6.12)
$$K^{uu}(z) \equiv [B^u(z^{-1}) D_2^{uu}(z)]_+ \cdot$$

When $K^{uu}(z)$ is singular at isolated points inside the unit circle, it follows from the Wold Decomposition Theorem that there exists a matrix $K^{uu*}(z)$ that is nonsingular inside the unit circle and satisfies:

(6.13)
$$K^{uu}[\exp(-i\theta)]\,K^{uu}[\exp(i\theta)]' = K^{uu^*}[\exp(-i\theta)]\,K^{uu^*}[\exp(i\theta)]'$$

almost everywhere (see also Section 3). Define

$$D_2^{uu^*}(z) \equiv [B^u(z^{-1})^{-1}\,K^{uu^*}(z)]_+ \,. \tag{6.14}$$

Then, D_2^* is observationally equivalent to D_2, where D_2^* is constructed by replacing $D_2^{uu}(z)$ by $D_2^{uu^*}(z)$. However, it is not necessarily true that

$$D_2^{uu}[\exp(-i\theta)]\,D_2^{uu}[\exp(i\theta)]' = D_2^{uu^*}[\exp(-i\theta)]\,D_2^{uu^*}[\exp(i\theta)]' \,, \tag{6.15}$$

so that the implied serial correlation properties for the unobservable component of y are different. Therefore, without additional restrictions, the spectral density function for the unobserved process y_2^u is not identified. Furthermore, $D^{uu^*}(z)$ may well have zeros inside the unit circle of the complex plane. In other words, it may require a moving-average representation for the unobserved process y_2^u which is *not* fundamental to construct a $D_2^*(z)$ associated with a fundamental moving-average representation for the observed process $[y_1', y_2^{o\prime}]'$. [8]

Conclusions

As indicated by the examples given in section two, the procedures described in this paper are applicable to a variety of linear rational expectations models. Therefore, the solution procedure, the characterization of identification, and the methods of estimation and inference for exact rational expectations models are useful tools for guiding a wide range of empirical applications. Chapter 6 of this volume contains an application.[9]

Notes

1. Actually, the assumption of stationarity of $m_t - m_{t-1}$ is not imposed in all of these papers. In Section 4 we see how to transform the model to accommodate unit roots in the process y.

2. This formulation does not include any explicit adjustment for the riskiness of the payout process. Such an adjustment, when appropriate, must be incorporated into a constant term or asset payouts

and values must be scaled appropriately by a process of used price contingent claims. Alternatively, Campbell and Shiller (1988) propose log-linear approximations as a means of incorporating risk-adjustments.

3. The probability space underlying this mean-square convergence has $[-\pi, \pi]$ as the collection of sample points, the Borel measurable subsets of $[-\pi, \pi]$ as a sigma algebra, and Lebesgue measure scaled by $1/2\pi$ as a probability measure.

4. The matrix of U in (3.30) is not identified even when (3.30) is imposed; however, the information structure associated with these $C(z)$'s is identified. As shown by Campbell and Shiller (1988), the present-value model (Example 2 in Section 2) implies a linear restriction on a finite-order vector autoregressive representation for y. This provides a convenient parameterization which enforces (3.30).

5. The rearranging of terms in the infinite sum in order to obtain (4.3) is justified given our assumptions about $B(z)$ and y.

6. This representation of the restrictions is superior to that used by Sargent (1979). Although the present restrictions are nonlinear in the coefficients of $D_d(z)$, they are considerably easier to compute than the restrictions that Sargent (1979) imposed directly on the vector autoregressive representation.

7. Time domain approximations to the likelihood function that assume $C(z)$ is nonsingular inside and on the unit circle of the complex plane may be inconvenient for many parameterizations of $D(z)$ because it may be difficult to restrict the parameterization of $D(z)$ in such a way as to ensure that $C(z)$ is nonsingular.

8. The working paper form of this chapter (Hansen and Sargent 1981e) contained an application to the term structure of interest rates.

9. A very similar phenomenon occurs in Futia (1981). In Futia's analysis two alternative rational expectations equilibria are compared. One endows economic agents directly with observations on an exogenous forcing process and the other presumes that information about this process can only be extracted from observations on endogenously determined prices. Futia showed that when endogenous information is sufficient to reveal the exogenous information, the two equilibria coincide. Otherwise the second equilibrium fails to exist.

4

Two Difficulties in Interpreting Vector Autoregressions

by Lars Peter HANSEN and Thomas J. SARGENT

Introduction

The equilibrium of a typical dynamic rational expectations model is a covariance stationary $(n \times 1)$ vector stochastic process $z(t)$. This stochastic process determines the manner in which random shocks to the environment impinge over time on agents' decisions and ultimately upon market prices and quantities. *Surprises*, i.e., random shocks to agents' information sets, prompt revisions in their contingency plans, thereby impinging on equilibrium prices and quantities.

Every $(n \times 1)$ covariance stationary stochastic process $z(t)$ can be represented in the form of a vector autoregression (of any finite order). Consequently, it is natural to represent the equilibrium of a dynamic rational expectations model in terms of its vector autoregression. A vector autoregression induces a vector of innovations which yields characterizations of the vector stochastic process via the "innovation accounting" techniques invented by Sims (1980).

In interpreting these innovation accountings, it is useful to understand the connections between the innovations recovered by vector autoregressions, on the one hand, and the random shocks to agents' information sets, on the other hand. From the viewpoint of interpreting vector autoregressions that are estimated without imposing restrictions from formal economic theories, it would be desirable if the innovations recovered by a vector autoregression could generally be expected to equal either the random shocks to agents' information sets, or else some simply interpretable functions of these random shocks. This paper describes two important settings in which no such simple connections exist. In these settings, without explicitly imposing the restrictions implied by the economic theory, it is impossible to make correct inferences about the shocks impinging on agents' information sets. In addition to describing these situations, we briefly indicate in each case how the economic theory can be used to deduce correct inferences about the shocks impinging on agents' information sets.

Let $z(t)$ be an $(n \times 1)$ vector, covariance stationary stochastic process. Imagine that $z(t)$ is observed at discrete points in time separated by the sampling interval Δ. A *vector autoregression* is defined by the projection equation

$$z(t) = \sum_{j=1}^{\infty} A_j^{\Delta} z(t - \Delta j) + a(t), \quad t = 0, \pm\Delta, \pm 2\Delta, \ldots \tag{1}$$

where $a(t)$ is an $(n \times 1)$ vector of population residuals from the regression with $Ea(t)a(t)^T = V$, and where the A_j^{Δ}'s are $(n \times n)$ matrices that, in general, are uniquely determined by the orthogonality conditions (or normal equations)

$$E\, z(t - \Delta j) a(t)^T = 0, \qquad j \geq 1\,. \tag{2}$$

The A_j^{Δ}'s in general are "square summable," that is, they satisfy

$$\sum_{j=1}^{\infty} \text{trace } A_j^{\Delta}\ A_j^{\Delta T} < +\infty \tag{3}$$

where the superscript T denotes matrix transposition. Equations (1)–(3) imply two important properties of $a(t)$. First, (1) and (2) imply that

$$E\ a(t)a(t - \Delta j)^T = 0, \qquad j \neq 0\,,$$

so that $a(t)$ is a vector white noise. Second, (1) and (3) imply that $a(t)$ is in the closed linear space spanned by $\{z(t),\ z(t-\Delta),\ z(t-2\Delta), \ldots\}$. Further, by successively eliminating all lagged $z(t)$'s from (1), we obtain the *vector moving-average representation*

$$z(t) = \sum_{j=0}^{\infty} C_j^{\Delta}\ a(t - \Delta j) \tag{4}$$

where the C_j^{Δ}'s are $n \times n$ matrices that satisfy

$$-\sum_{j=0}^{\infty} C_j^{\Delta}\ A_{s-j}^{\Delta} = \begin{cases} I & s = 0 \\ 0 & s \neq 0 \end{cases}$$

where $A_0^\Delta \equiv -I$. The C_j^Δ's in (4) satisfy

$$\sum_{j=0}^{\infty} \text{trace } C_j^\Delta \, C_j^{\Delta T} < +\infty \, . \tag{5}$$

Equations (4) and (5) imply that $z(t)$ is in the closed linear space spanned by $(a(t),\ a(t-\Delta),\ a(t-2\Delta),\ \ldots)$. Thus, the closed linear space spanned by $(z(t),\ z(t-\Delta),\ \ldots)$ equals the closed linear space spanned by $(a(t),\ a(t-\Delta),\ \ldots)$. In effect, $a(t)$ is a stochastic process that forms an orthogonal basis for the stochastic process $z(t)$, and which can be constructed from $z(t)$ via a *Gram-Schmidt* process. The property of the vector white noise $a(t)$ that it is contained in the linear space spanned by current and lagged $z(t)$'s is said to mean that "$a(t)$ is a *fundamental* white noise for the $z(t)$ process."

It is a moving-average representation for $z(t)$ in terms of a fundamental white noise that is automatically recovered by vector autoregression.[1] However, there are in addition a variety of *other* moving-average representations for $z(t)$ of the form

$$z(t) = \sum_{j=0}^{\infty} \tilde{C}_j^\Delta \, \tilde{a}(t-\Delta j) \tag{6}$$

where $\tilde{a}(t)$ is an $(n \times 1)$ vector white noise in which the linear space spanned by $(\tilde{a}(t), \tilde{a}(t-\Delta),\ \ldots)$ is strictly *larger* than the linear space spanned by current and lagged $z(t)$'s. Current and lagged $z(t)$'s fail to be "fully revealing" about the $\tilde{a}(t)$'s in such representations.

Representation (4) induces the following decomposition of j-step ahead prediction errors[2]

$$\begin{aligned} &E(z(t) - \hat{E}_{t-j} z(t)) \, (z(t) - \hat{E}_{t-j} z(t))^T \\ &= \sum_{k=0}^{j-1} C_k^\Delta \, V \, C_k^{\Delta T} \, . \end{aligned} \tag{7}$$

By studying versions[3] of decomposition (7), Sims has shown how the j-step ahead prediction error variance can be decomposed into parts attributable to *innovations* in particular components of the vector $z(t)$.

Sims has described methods for estimating vector autoregressions and for obtaining alternative fundamental moving-average representations. He has also created a useful method known as "innovation accounting" that is based on decomposition (7). In the hands of Sims and other skilled analysts, these methods have been used successfully to detect interesting patterns in data, and to suggest possible interpretations of them in terms of the responses of systems of people to surprise events.

This paper focuses on the question of whether dynamic economic theories readily appear in the form of a fundamental moving-average representation (4), so that the vector white noises $a(t)$ recovered by vector autoregressions are potentially interpretable in terms of the white noise impinging on the information sets of the agents imagined to populate the economic model. This question is important because it influences the ease with which one can interpret the variance decompositions (or innovation accounts) and the responses to innovations $a(t)$ that are associated with the fundamental moving average (4).

This paper is organized as follows. Section 1 describes a class of discrete-time models whose equilibria can be represented in the form

$$z_t = \sum_{j=0}^{\infty} D_j \varepsilon_{t-j} \tag{8}$$

where ε_t is an $(n \times 1)$ vector white noise; D_j is an $(n \times n)$ matrix for each j; and $\sum_{j=0}^{\infty}$ trace $D_j D_j^T < \infty$. Here ε_t represents a set of shocks to agents' information sets. We study how the ε_t of representation (8) are related to the $a(t)$ of the (Wold) representation (4), and how the D_j's of (8) are related to the C_j^{Δ}'s of (4). We describe contexts in which $a(t)$ fails to match up with ε_t and C_j^{Δ} fails to match up with D_j because $z(t)$ fails to be fully revealing about ε_t. Such examples were encountered earlier by Hansen and Sargent (1980a), Futia (1981), and Townsend (1983). The discussion in Section 1 assumes that the sampling interval Δ equals the sampling interval in terms of which the economic model is correctly specified.

Section 2 describes a class of continuous time models whose equilibria are represented in the form

$$z_t = \int_0^{\infty} p(\tau) w(t-\tau) d\tau \tag{9}$$

where $w(t)$ is an m-dimensional continuous time white noise and $p(\tau)$ is an $(n \times m)$ function satisfying $\int_0^{\infty}$ trace $p(\tau)p(\tau)^T d\tau < +\infty$. In (9), $w(t)$ represents shocks to agents' information set. It is supposed that economic decisions occur in continuous time according to (9), but that the econometrician possesses data only at discrete intervals of time.

Section 2 studies the relationship between the $w(t)$ of (9) and the $a(t)$ of (4), and also the relationship between $p(\tau)$ of (9) and the C_j^{Δ} of (4). In general, these pairs of objects do not match up in ways that can be determined without the imposition of restrictions from a dynamic economic theory.[4]

1. Unrevealing Stochastic Processes

We consider a class of discrete time linear rational expectations models that can be represented as the solution of the following pair of stochastic difference equations

$$\begin{aligned} H(L)y_t &= E_t\, J(L^{-1})^{-1} p x_t \\ x_t &= K(L)\, \varepsilon_t \end{aligned} \tag{1.1}$$

where

$$\begin{aligned} H(L) &= H_0 + H_1 L + \ldots + H_{m_1} L^{m_1} \\ J(L) &= J_0 + J_1 L + \ldots + J_{m_2} L^{m_2} \\ K(L) &= \sum_{j=0}^{\infty} K_j L^j, \qquad K_0 = I \\ \varepsilon_t &= x_t - E(x_t \mid x_{t-1},\, x_{t-2},\, \ldots) \end{aligned} \tag{1.2}$$

In (1.1), y_t is an $n_1 \times 1$ vector, while x_t is an $n_2 \times 1$ vector. In (1.2), J_j and H_j are $(n_1 \times n_1)$ matrices, while K_j is an $(n_2 \times n_2)$ matrix. In (1.1), p is an $(n_1 \times n_2)$ matrix. We assume that the zeroes of $\det H(z)$ lie outside the unit circle, that those of $\det J(z)$ lie inside the unit circle, and that those of $\det K(z)$ do not lie outside the unit circle.

Many discrete time linear rational expectations models are special cases of (1.1). For example, interrelated factor demand versions of Lucas-Prescott equilibrium models are special cases with $J(L^{-1}) = H(L^{-1})^T$ and with $H(L^{-1})^T\, H(L)$ being the matrix factorization of the Euler equation that is solved by the fictitious social planner (see Hansen and Sargent (1981a) and Eichenbaum (1983) for some examples). Kydland-Prescott equilibria with feedback from market-wide variables to forcing variables that individual agents face parametrically form a class of examples with $H(L^{-1})^T \neq J(L^{-1})$ (see Sargent 1981). Other examples with $H(L^{-1})^T \neq J(L^{-1})$ arise in the context of various dominant player equilibria of linear quadratic differential games (see Hansen, Epple, and Roberds 1985). Finally, market equilibrium models of the Kennan (1988)-Sargent (1987b) variety, an example of which is studied below, solve a version of (1.1) with $H(L^{-1})^T \neq J(L)$. Models of this general class are studied by Whiteman (1983).

Hansen and Sargent (1981a) displayed a convenient representation of the solution of models related but not identical to (1.1). To adapt their results, first obtain the partial fractions representation of $J(z^{-1})^{-1}$. We have $J(z^{-1})^{-1} = \text{adj } J(z^{-1})/ \det J(z^{-1})$. Let

$$\det J(z^{-1}) = \lambda_0(1 - \lambda_1 z^{-1}) \; \ldots \; (1 - \lambda_k z^{-1})$$

where $k = m_2 \cdot n_1$ and $|\lambda_j| < 1$ for $j = 1, \ldots, k$. The λ_j's are the zeroes of $\det J(z^{-1})$ which are assumed to be distinct. Then we have

$$J(z^{-1})^{-1} = \sum_{j=1}^{k} \frac{M_j}{(1 - \lambda_j z^{-1})} \tag{1.3}$$

where

$$M_j = \lim_{z \to \lambda_j} \; J(z^{-1})^{-1} \, (1 - \lambda_j z^{-1}) \, . \tag{1.4}$$

Substitute (1.3) into (1.1) to obtain

$$H(L)\, y_t = E_t \sum_{j=1}^{k} \frac{M_j}{1 - \lambda_j L^{-1}} \, p\, x_t \tag{1.5}$$

Hansen and Sargent (1980a) establish that

$$E_t \, \frac{M_j}{1 - \lambda_j L^{-1}} \, p \, x_t = M_j \, p \left(\frac{L K(L) - \lambda_j K(\lambda_j)}{L - \lambda_j} \right) \varepsilon_t \, . \tag{1.6}$$

Define the operator M by

$$M(K(L)) = \sum_{j=1}^{k} M_j \, p \left(\frac{L K(L) - \lambda_j K(\lambda_j)}{L - \lambda_j} \right) . \tag{1.7}$$

Then, using (1.5), (1.6), and (1.7) we have the representation of the solution

$$\begin{aligned} H(L) \, y_t &= M(K(L)) \, \varepsilon_t \\ x_t &= K(L) \, \varepsilon_t \end{aligned} \tag{1.8}$$

A vector stochastic process (y_t^T, x_t^T) governed by (1.8) generally has a singular spectral density at all frequencies because (y_t^T, x_t^T) consists of $n_1 + n_2$ variables being driven by only n_2 white noises. Such a model implies that various of the first n_1 equations of the following model, which is equivalent to (1.8),[5]

$$\begin{aligned} H(L)\, y_t &= M(K(L))\; K(L)^{-1}\, x_t \\ x_t &= K(L)\; \varepsilon_t \end{aligned}$$

will fit perfectly (i.e., possess sample $\bar{R}^2$'s of 1).

To avoid this implication of no errors in various of the equations of the model, while still retaining the model, one path that has been suggested is to assume that the econometrician seeks to estimate (1.8), but that he possesses data only on a subset of the variables in $(y_t,\, x_t)$. (See Hansen and Sargent 1980a). One common procedure, but not the only one possible, is the following one described by Hansen and Sargent (1980a). Assume that (1.8) holds, but that the econometrician only has data on a subset of observations x_{2t} of x_t. Further suppose that the second equation of (1.1) can be partitioned and restricted as

$$x_t = \begin{pmatrix} x_{1t} \\ x_{2t} \end{pmatrix} = \begin{pmatrix} K_1(L) & 0 \\ 0 & K_2(L) \end{pmatrix} \begin{pmatrix} \varepsilon_{1t} \\ \varepsilon_{2t} \end{pmatrix}. \tag{1.9}$$

Then (1.8) assumes a special form which can be represented as

$$\begin{pmatrix} H(L) & 0 \\ 0 & I \end{pmatrix} \begin{pmatrix} y_t \\ x_{2t} \end{pmatrix} = \begin{pmatrix} M(K_1(L)) & M(K_2(L)) \\ 0 & K_2(L) \end{pmatrix} \begin{pmatrix} \varepsilon_{1t} \\ \varepsilon_{2t} \end{pmatrix}. \tag{1.10}$$

The idea is to imagine that the econometrician is short of observations on a sufficient number of series, those forming x_{1t}, to make the $(y_t^T,\, x_{2t}^T)$ process described by (1.10) have a nonsingular spectral density matrix at all frequencies. To accomplish this, it will generally be sufficient that the dimension of the vector of variables of $(y_t^T,\, x_{2t}^T)$ be less than or equal to the dimension of $(\varepsilon_{1t}^T,\, \varepsilon_{2t}^T)$. For the argument below, we will consider the case often encountered in practice in which $(y_t^T,\, x_{2t}^T)$ and $(\varepsilon_{1t}^T,\, \varepsilon_{2t}^T)$ have equal dimensions. Thus we assume that x_{1t} is an $(n_1 \times 1)$ vector, so that ε_{1t} is an $(n_1 \times 1)$ vector of white noises.[6]

Equation (1.10) implies the moving-average representation for $(y_t,\, x_{2t})$

$$\begin{pmatrix} y_t \\ x_{2t} \end{pmatrix} = \begin{pmatrix} H(L)^{-1}M(K_1(L)) & H(L)^{-1}M(K_2(L)) \\ 0 & K_2(L) \end{pmatrix} \begin{pmatrix} \varepsilon_{1t} \\ \varepsilon_{2t} \end{pmatrix}. \tag{1.11}$$

Equation (1.11) is a moving average that expresses $(y_t,\, x_{2t})$ in terms of current and lagged values of the white noises $(\varepsilon_{1t},\, \varepsilon_{2t})$ that are the innovations in the information sets (x_{1t}, x_{2t}) of the agents in the model. Equivalently, $(\varepsilon_{1t}, \varepsilon_{2t})$ are *fundamental* for x_{1t}, x_{2t}, the one step-ahead errors in predicting x_{1t}, x_{2t} from their own pasts being expressible as linear combinations of ε_{1t}, ε_{2t}.

Granted that the linear space spanned by current and lagged $(\varepsilon_{1t}, \varepsilon_{2t})$ equals that spanned by current and lagged values of the agents' information (x_{1t}, x_{2t}), there remains the question of whether this space equals that spanned by current and lagged values of the econometrician's information (y_t, x_{2t}). As is evident from the construction of (1.11), the latter space is included in the former. The question is whether they are equal. This question is an important one from the viewpoint of interpreting vector autoregressions because a vector autoregression by construction would recover a vector moving average for (y_t, x_{2t}) that is driven by a vector white noise a_t that is fundamental for (y_t, x_{2t}), i.e., one that is in the linear space spanned by current and lagged values of (y_t, x_{2t}). If this space is smaller than the one spanned by current and lagged values of agents' information $(\varepsilon_{1t}, \varepsilon_{2t})$, then the moving-average representation recovered by the vector autoregression will in general give a distorted impression of the response of the system to surprises from agents' viewpoint.

The vector white noise $(\varepsilon_{1t}, \varepsilon_{2t})$ is fundamental for (y_t, x_{2t}) if and only if the zeroes of

$$\det \begin{pmatrix} H(z)^{-1}M(K_1(z)) & H(z)^{-1}M(K_2(z)) \\ 0 & K_2(z) \end{pmatrix}$$
$$= \det H(z)^{-1} \cdot \det M(K_1(z)) \cdot \det K_2(z)$$

do not lie inside the unit circle. The zeroes of $\det K_2(z)$ do not lie inside the unit circle by assumption, and $\det H(z)^{-1} = 1/\det H(z)$ is a function with all its poles outside the unit circle. Therefore, the necessary and sufficient condition that $(\varepsilon_{1t}, \varepsilon_{2t})$ be fundamental for (y_t, x_{2t}) is that

$$\det M(K_1(z)) = 0 \Rightarrow |z| \geq 1 , \tag{1.12}$$

or equivalently, using (1.7),

$$\det \sum_{j=1}^{k} M_j p \left(\frac{zK_1(z) - \lambda_j K_1(\lambda_j)}{z - \lambda_j} \right) = 0 \Rightarrow |z| \geq 1 . \tag{1.12$'$}$$

In general, condition (1.12) is not satisfied. For some specifications of $K_1(L)$ and $J(L^{-1})$, which determines the (M_j, λ_j) via (1.3)–(1.4), condition (1.12) is met, while for others, it is not met. Hansen and Sargent (1980a) encountered a class of examples where (1.12) is not met. Furthermore, the class of cases for which (1.12) fails to be met is not thin in any natural sense. Our conclusion is that for the class of models defined by (1.1)–(1.2), the moving-average representation (1.11) that is expressed in terms of the white noises that are fundamental for agents' information sets in general cannot be expected to be fundamental for the econometrician's data set (y_t, x_{2t}). Equivalently, current and lagged values of (y_t, x_{2t}) fail to be fully revealing of current and lagged values of $(\varepsilon_{1t}, \varepsilon_{2t})$.

For convenience, let us rewrite (1.10) as

$$S(L)\, z_t = R(L)\, \varepsilon_t \tag{1.13}$$

where

$$S(L) = \begin{pmatrix} H(L) & 0 \\ 0 & I \end{pmatrix}, \quad R(L) = \begin{pmatrix} M(K_1(L)) & M(K_2(L)) \\ 0 & K_2(L) \end{pmatrix}$$

$$\varepsilon_t = \begin{pmatrix} \varepsilon_{1t} \\ \varepsilon_{2t} \end{pmatrix}, \quad z_t = \begin{pmatrix} y_t \\ x_{2t} \end{pmatrix} .$$

The condition that ε_t be fundamental for z_t is then expressible as the condition that the zeroes of det $R(z)$ not lie inside the unit circle. If this condition is violated, then a Wold representation for z_t, which is what is recovered via vector autoregression, will be related to representation (1.13) as follows. It is possible to show[7] that there exists a matrix polynomial $G(L)^T = \sum_{j=0}^{\infty} G_j^T L^j$ with the following properties:

(*i*) $G(L^{-1})\, G(L)^T = I$

(*ii*) $G(L)^T$ is one-sided in nonnegative powers of L.

(*iii*) $R(z)\, G(z^{-1})$ has a power series expansion with square summable coefficients and the zeroes of det $(R(z)\, G(z^{-1}))$ are not inside the unit circle of the complex plane.

Evidently, we can represent $R(L)\,\varepsilon_t$ as

$$R(L)\varepsilon_t = R(L)\, G(L^{-1})\, G(L)^T\, \varepsilon_t$$

or

$$R(L)\varepsilon_t = R^*(L)\varepsilon_t^* , \tag{1.14}$$

where

$$R^*(L) = R(L)\, G(L^{-1}) \tag{1.15}$$

and

$$\varepsilon_t^* = G(L)^T \varepsilon_t \,. \tag{1.16}$$

Then a Wold representation for z_t corresponding to (1.13) is

$$S(L) z_t = R^*(L) \varepsilon_t^* \,. \tag{1.17}$$

By construction, the ε_t^* process defined in (1.16) is a vector white noise that lies in the space spanned by current and past values of the process z_t, in other words ε_t^* is *fundamental* for z_t. It is property *(iii)* which assures that the spanning condition is satisfied. If, in addition, $R^*(L)$ has an inverse that is one-sided in nonnegative powers of L, we can obtain a representation for ε_t in terms of current and past z_t's:

$$\varepsilon_t^* = R^*(L)^{-1} \, S(L) \, z_t \,. \tag{1.18}$$

In general, $R^*(L)$ given by (1.15) can be very different from $R(L)$, so that the impulse response of the system to ε_t (which are the innovations to *agents*' information set) can look very different from the impulse response to ε_t^* (which are the innovations to the *econometrician's* information set). Equation (1.16) and property *(ii)* above shows that ε_t^* is a one-sided distributed lag of current and past ε_t's so that the innovation ε_t^* reflects information that is *old news* to the agents in the model at t. Only in the special case that none of the zeros of det $R(z)$ are inside the unit circle, so that $G(L)^T$ can be taken to be the identity operator, does ε_t^* given by (1.16) only respond to contemporary news ε_t possessed by the agents. In the case that $G(L)^T \neq I$, it is generally true that[8]

$$E(R_0^* \varepsilon_t^*) (R_0^* \varepsilon_t^*)^T \; > E(R_0 \varepsilon_t) (R_0 \varepsilon_t)^T \,. \tag{1.19}$$

This statement that the contemporaneous covariance matrix of $R_0^* \, \varepsilon_t^*$ is larger than that of $R_0 \, \varepsilon_t$ concisely summarizes how ε_t^* contains less information about the z_t process than does ε_t.

We now describe a concrete hypothetical numerical example, one in which the econometrician observes no x's, only y's, so that (1.10) takes the special form

$$H(L) \, y_t = M(C_1(L)) \, \varepsilon_{1t} \,. \tag{1.20}$$

The model is one of the dynamics of price and quantity in a single market, and is related to ones studied by Sargent (1987b) and Kennan

(1988). Behavior by agents on the two sides of the market, supply and demand, are each described by linear Euler equations. The Euler equation for suppliers is

$$-E_t\left\{[h_s + g_s(1-\beta L^{-1})\,(1-L)]\,q_t\right\} + p_t = s_t \tag{1.21}$$

where p_t is price at time t, q_t is quantity supplied at time t, s_t is a supply stock at t, β is a discount factor between zero and one, and h_s and g_s are parameters of the suppliers' cost function. Equation (1.21) is typical of the kind of Euler equation that characterizes the optimum problem of a competitive firm facing adjustment costs that are quadratic in $(1-L)\,q_t$.

The Euler equation for demanders is

$$-E_t\left\{h_d + g_d[a(\beta L^{-1})\,a(L)]\right\}q_t - p_t = d_t \tag{1.22}$$

where $a(L) = a_0 + a_1 L + a_2 L^2 + a_3 L^3 + a_4 L^4$. In (1.22), h_d and g_d are preference parameters, while the parameters $[a_0, a_1, a_2, a_3, a_4]$ characterize a technology by which purchases of q_t at t give rise to utility-generating services to the demander in subsequent periods. In (1.22), d_t is a stochastic process of disturbances to demand.

To complete the model, we specify the stochastic law of motion for the forcing processes, i.e., the demand and supply shocks. These shocks are assumed to satisfy

$$\begin{aligned} s_t &= B_s(L)w_{st} \\ d_t &= B_d(L)w_{dt} \end{aligned} \tag{1.23}$$

where $B_s(z)$ and $B_d(z)$ are scalar polynomials with zeros that are outside the unit circle. The w_{st} and w_{dt} processes are mutually uncorrelated white noises so that w_{st} is the innovation in the supply shock, s_t, and w_{dt} is the innovation in the demand shock, d_t. Economic agents are assumed to observe current and past values of both shocks and hence also the innovations in both shocks.

The typical supplier and demander are both assumed to view the stochastic processes p_t, d_t, s_t as beyond their control and to choose a stochastic process for q_t. At time t, both suppliers and demanders have the common information set of $\{p_r, d_r, s_r, q_{r-1}\ ;\ r \leq t\}$. Both suppliers and demanders choose contingency plans for q_t as a function of this information set. As discussed in Kennan (1988) and Sargent (1987b), an *equilibrium* is a stationary stochastic process for $\{q_t,\ p_t\}$ that solves the pair of difference equations (1.21)–(1.22).[9]

This model fits into our general set up (1.1) as follows. Let

$$y_t = \begin{pmatrix} q_t \\ p_t \end{pmatrix}, \quad \varepsilon_t = \begin{pmatrix} w_{st} \\ w_{dt} \end{pmatrix} .$$

$$K_2(L) = 0, \qquad x_{2t} \equiv 0 .$$

$$K_1(L) = \begin{pmatrix} B_s(L) & 0 \\ 0 & B_d(L) \end{pmatrix} .$$

Define a matrix polynomial

$$E(L) = \begin{pmatrix} -(h_s + g_s(1-L)(1-\beta L^{-1})) & 1 \\ -(h_d + g_d a(L) a(\beta L^{-1})) & -1 \end{pmatrix} .$$

Then polynomial matrices $H(L)$ and $J(L)$ that are one-sided in non-negative powers of L can be found such that

$$J(L^{-1})\, H(L) = E(L)$$

and such that the zeroes of $\det J(z)$ lie inside the unit circle, while the zeroes of $\det H(z)$ lie outside the unit circle. (See Whiteman 1983, and Gohberg, Lancaster, and Rodman 1982 for proofs of the existence of such a matrix factorization, and for descriptions of algorithms for achieving the factorization.)[10]

In this model, it is possible for the demand and supply shocks to generate an information set that is strictly larger than that generated by current and past quantities and prices. So an econometrician using innovation accounts derived from observations on quantities and prices may not obtain innovations that are linear combinations of the contemporaneous innovations to the demand and supply shocks.

The equilibrium of the model has representation

$$S(L) \begin{pmatrix} q_t \\ p_t \end{pmatrix} = R(L) \begin{pmatrix} w_{dt} \\ w_{st} \end{pmatrix} \tag{1.24}$$

where $S(L) = H(L)$ and $R(L) = M(C_1(L))$, and where $S(L)$ is a (2×2) fourth-order matrix polynomial in L, with the zeroes of $\det S(z)$ outside the unit circle. The zeroes of $\det R(z)$ can be on either side of the unit circle in this example. Only when the zeroes of $\det R(z)$ are not inside the unit circle can the one-step ahead forecast errors from the vector autoregression of prices and quantities be expressed as linear combinations of the contemporaneous demand and supply shock innovations $(w_{st},\, w_{dt})$.

We have computed a numerical example that illustrates the ideas discussed above.[11] We set the parameters of the model as follows:[12]

$$h_s = h_d = 1\,,\ g_s = 10,\ g_d = .1$$
$$a(L) = 1 + .8L + .6L^2 + .4L^3 + .2L^4$$
$$B_d(L) = (1 + .6L)\ (1 + .4L)\ (1 + .2L)$$
$$B_s(L) = (1 - .8L)\ (1 + .4L)\ (1 + .2L)$$
$$\beta = 1/1.05$$
$$E\,w_{st}^2 = .5\,,\ E\,w_{dt}^2 = 4\,,\ E\,w_{st}\,w_{dt} = 0$$

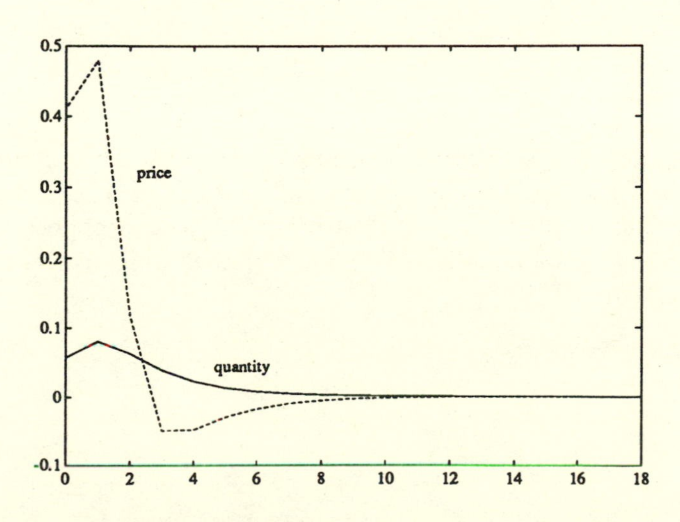

Figure 1a. Impulse response functions to first innovation in $S(L)\begin{bmatrix} q_t \\ p_t \end{bmatrix} = R(L)\epsilon_t$.

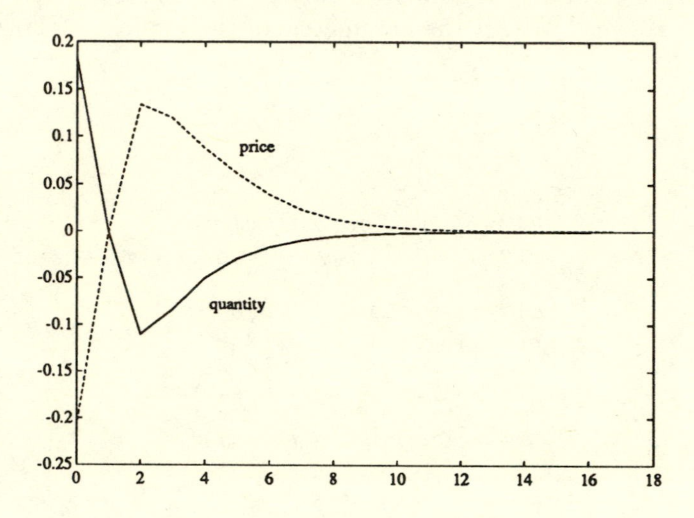

Figure 1b. Impulse response functions to second innovation in $S(L)\begin{bmatrix} q_t \\ p_t \end{bmatrix} = R(L)\varepsilon_t$.

Figure 1 displays response functions of (q_t, p_t) to an impulse in the innovation (w_{dt}, w_{st}) to agents' information set. This corresponds to representation (1.13) or (1.24). Figure 2 displays the response function of (q_t, p_t) to a fundamental (Wold) innovation ε_t^* which corresponds to representation (1.17) or

$$S(L)\begin{bmatrix} q_t \\ p_t \end{bmatrix} = R^*(L)\,\varepsilon_t^* \,. \tag{1.25}$$

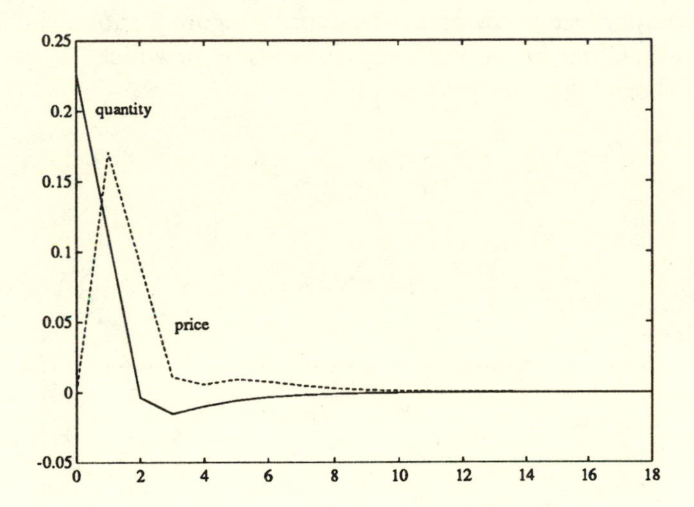

Figure 2a. Impulse response functions to first innovation in $S(L)\begin{bmatrix} q_t \\ p_t \end{bmatrix} = R^*(L)\varepsilon_t^*$. Quantity is 'first' in the orthogonalization.

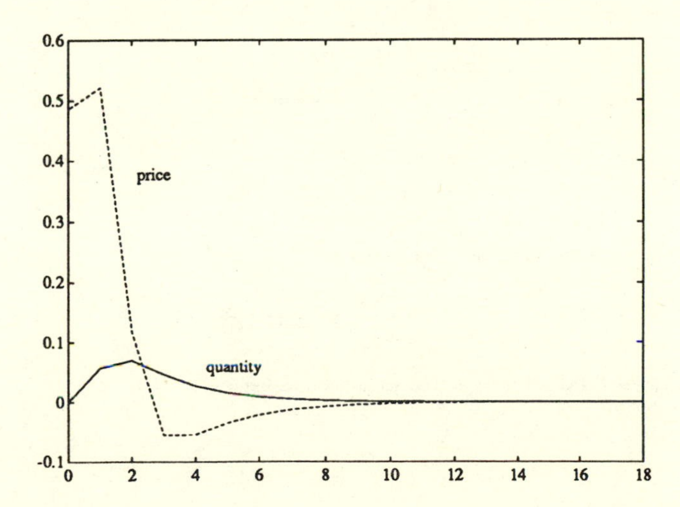

Figure 2b. Impulse response functions to second innovation in $S(L)\begin{bmatrix} q_t \\ p_t \end{bmatrix} = R^*(L)\varepsilon_t^*$. Quantity is 'first' in the orthogonalization.

In Figure 2, we have followed Sims (1980) in normalizing R_0^* and ε_t^* by selecting a version of ε_t^* which has diagonal contemporaneous covariance matrix, and for which the variance of ε_{1t}^* is maximal. This amounts to "letting q_t go first" in a Gram-Schmidt procedure that orthogonalizes the contemporaneous covariance matrix.[13] Figure 3 shows the impulse response functions in the Wold representation for which "p_t goes first" in the orthogonalization procedure.

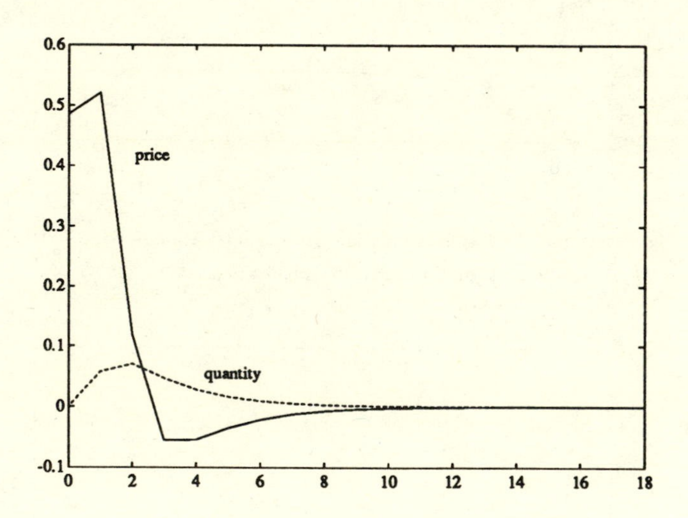

Figure 3a. Impulse response functions for

$$S(L)\begin{bmatrix} q_t \\ p_t \end{bmatrix} = R^*(L)UU'\varepsilon_t^*,$$

$UU' = I$. **Price is 'first' in the orthogonalization.**

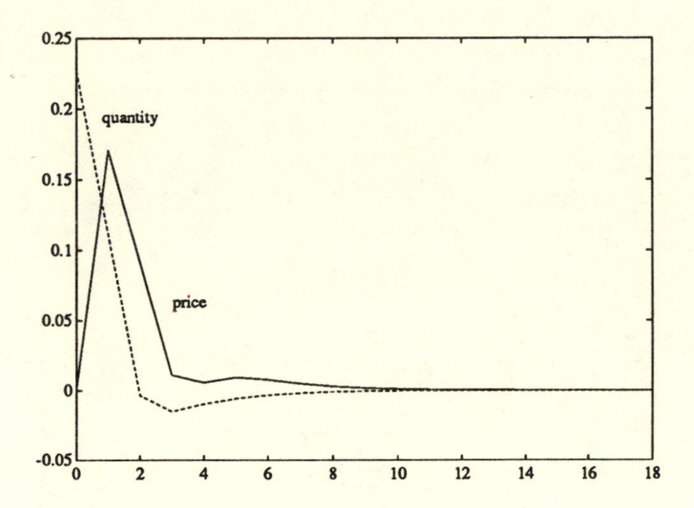

Figure 3b. Impulse response functions for

$$S(L)\begin{bmatrix} q_t \\ p_t \end{bmatrix} = R^*(L)UU'\varepsilon_t^*,$$

$UU' = I$. **Price is 'first' in the orthogonalization.**

The representation in Figure 3 is related to that in Figure 2 via the equation

$$S(L)\begin{bmatrix} q_t \\ p_t \end{bmatrix} = (R^*(L)\, U)\,(U^T\, \varepsilon_t^*)\,,$$

where U is an orthogonal matrix that implements the Gram-Schmidt procedure that puts p_t first in the orthogonalization process.

That Figures 2 and 3 are very similar is a reflection of the fact that the contemporaneous covariance matrix of $R_0^*\varepsilon_t^*$ is nearly diagonal:

$$E\,(R_0^*\,\varepsilon_t^*)\,(R_0^*\,\varepsilon_t^*)^T = \begin{bmatrix} .0517 & -.0332 \\ -.0337 & .2354 \end{bmatrix}$$

From representation (1.24) corresponding to Figure 1, we computed

$$E\,(R_0\,\varepsilon_t)\,(R_0\,\varepsilon_t)^T = \begin{bmatrix} .0374 & -.0149 \\ -.0149 & .2121 \end{bmatrix}$$

Thus, inequality (1.19) holds strictly for our example.

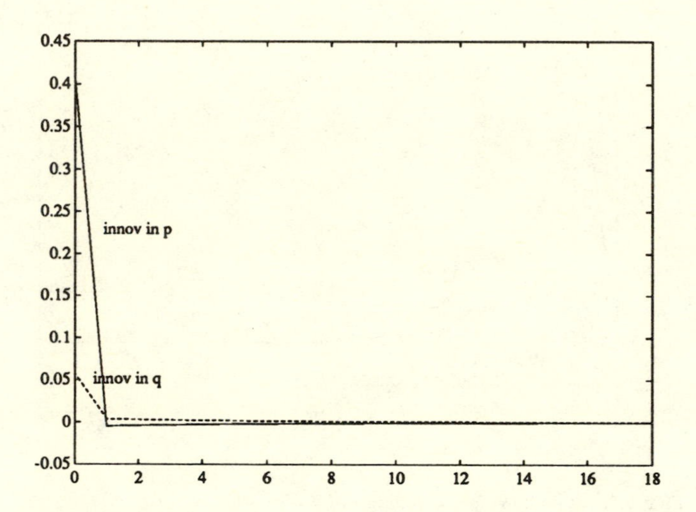

Figure 4a. Impulse response functions for $\varepsilon_t^* = G(L)'\varepsilon_t$. Response of ε_t^* to a demand innovation.

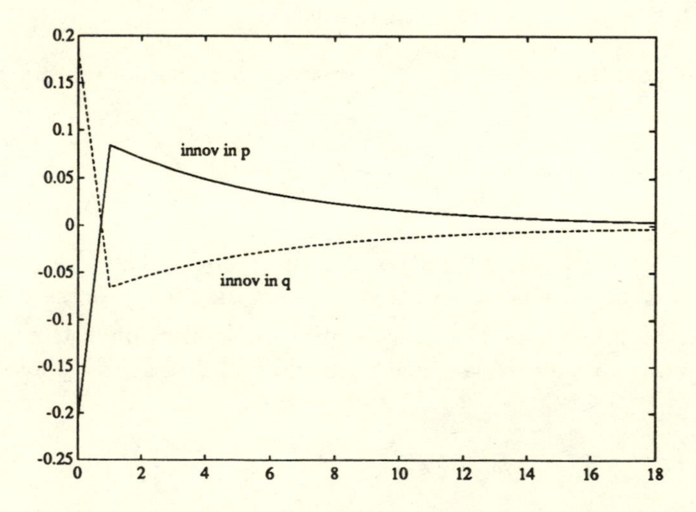

Figure 4b. Impulse response functions for $\varepsilon_t^* = G(L)'\varepsilon_t$. Response of ε_t^* to a supply innovation.

Figure 4 depicts the response of ε_t^* to the innovations (w_{dt}, w_{st}) in agents' information sets, which corresponds to representation (1.16), namely $\varepsilon_t^* = G(L)^T \varepsilon_t$. Notice that the demand innovation w_{dt} gives rise almost entirely to a contemporary response in both the innovation in p and the innovation in q, with the numerically much larger response being in the innovation to p. However, the supply innovation w_{st} gives rise to distributed responses in the innovations both to p and to q, with numerically the larger one being in the q innovation. Figure 4 is consistent with the approximation that the innovation in p from representation (1.24) reflects the innovation to demand in a timely manner, but that the innovation to q is a distributed lag mainly of the innovation in supply.

These interpretations bear up when we compare Figures 1 and 2. Notice how much Figure 1a resembles Figure 2b, which is consistent with the interpretation of the innovation in p as an innovation in the demand shock. However, Figure 1b does not very much resemble Figure 2a. Indeed, Figure 1b more closely resembles Figure 4b, which depicts the response of ε_t^* to the supply innovation. This is understandable in view of the nearly entirely contemporaneous response of q to its own innovation depicted in Figure 2a.

1a. Remedies in Discrete Time

The preceding difficulty can be circumvented if a sufficiently restrictive dynamic economic theory is imposed during estimation. Hansen and Sargent (1980a) describe methods for estimating $S(L)$ and $R(L)$ subject to extensive cross-equation restrictions of the rational expectations variety. The approach is to use the method of maximum likelihood to estimate free parameters of preferences and constraint sets, of which the parameters of $S(L)$ and $R(L)$ are in turn functions. These methods do not require that the zeroes of $\det R(z)$ be restricted, and in particular are capable of recovering good estimates of $R(z)$ even when some of the zeroes of $\det R(z)$ are inside the unit circle.

Simply take representation (1.13) and operate on both sides by the two-sided inverse $R(L)^{-1}$ to obtain

$$\varepsilon_t = R(L)^{-1} S(L) z_t \,. \tag{1.26}$$

When some of the zeros of $\det R(z)$ are inside the unit circle, $R(L)^{-1}$ is two-sided, so that (1.26) expresses ε_t as a linear function of past, present, and future z_t's. By using this equation together with estimates of the model's parameters, it would be possible to construct estimates of ε_t as functions of an observed record on $\{z_t\}$. Equation (1.26) once again illustrates how ε_t fails to lie in the linear space spanned by current and lagged z's.

2. Time Aggregation

Consider a linear economic model that is formulated in continuous time, and which can be represented as

$$z(t) = \int_0^\infty p(\tau)\, w(t-\tau) d\tau \tag{2.1}$$

where $z(t)$ is an $(n \times 1)$ vector stochastic process, $w(t)$ is an $(m \times 1)$ vector white noise with $Ew(t)w(t-s)^T = \delta(t-s)I$, δ is the Dirac delta generalized function, and $p(\tau)$ is an $(n \times m)$ matrix function that satisfies $\int_0^\infty$ trace $p(\tau)p(\tau)^T\, d\tau < +\infty$. We let $P(s) = \int_0^\infty e^{-s\tau} p(\tau) d\tau$, i.e., $P(s)$ is the Laplace transform of $p(\tau)$. Sometimes we shall find it convenient to write (2.1) in operator notation

$$z(t) = P(D)w(t) \tag{2.2}$$

where D is the derivative operator. We shall assume that det $P(s)$ has no zeroes in the right half of the complex plane. This guarantees that square integrable functionals of $(z(t-s),\ s \geq 0)$ and of $(w(t-s),\ s \geq 0)$ span the same linear space, and is equivalent to specifying that (2.1) is a Wold representation for $z(t)$.

A variety of continuous time stochastic linear rational expectations models have equilibria that assume the form of the representations (2.1) or (2.2). Hansen and Sargent (1981d) provide some examples. In these examples, the continuous time white noises $w(t)$ often have interpretations as innovations in the uncontrollable processes that agents care about forecasting, and which stochastically drive the model. These include processes that are imagined to be observable to both the econometrician and the private agent (e.g., various relative prices and quantities) and also those which are observable to private agents but are hidden from the econometrician (e.g., random disturbances to technologies, preferences, and maybe even particular factors of production such as *effort* or capital of specific kinds). The $w(t)$ process is economically interpretable as the continuous time innovation to private agents, because a forecast error of the variables in the model over any horizon $t+\tau$ which the private agents are assumed to make at t can be expressed as a weighted sum of $w(s)$, $t < s \leq t+\tau$. Thus, to private agents the $w(t)$ process represents *news* or *surprises*.

In rational expectations models, typically there are extensive restrictions across the rows of $P(D)$. In general these restrictions leave open the possibility that the current and lagged values of the $w(t)$ process span a larger linear space than do current and lagged values of the

$z(t)$ process. This outcome can possibly occur even if the dimension m of the $w(t)$ process is less than or equal to the dimension n of the $z(t)$ process. This is the continuous time version of the phenomenon that we treated for discrete time in the previous section. In the present section, we ignore this phenomenon, by assuming that det $P(s)$ has no zeroes in the right half of the complex plane.

For this continuous time specification, there exists a discrete time moving-average representation

$$z_t = C(L)a_t \tag{2.3}$$

where $C(L)$ is an infinite order, $(n \times n)$ polynomial in the lag operator L, where a_t is a vector white noise with $Ea_t a_t^T = W$, and where $a_t = z_t - \hat{E}[z_t|, z_{t-1}, \ldots]$. The operator $C(L)$ and the positive semi-definite matrix W solve the following equation, subject to the side condition that the zeroes of $\det C(z)$ do not lie inside the unit circle:[14]

$$C(e^{-i\omega})W\, C(e^{i\omega})^T = \sum_{j=-\infty}^{+\infty} P(i\omega + 2\pi ij)\, P(-i\omega - 2\pi ij)^T\ . \tag{2.4}$$

When z_t has a discrete time autoregressive representation, the discrete time innovations a_t are related to the $w(t)$ process by the formula

$$a_t = C(L)^{-1}\, P(D)\; w(t)$$

or

$$a_t = V(L)\; P(D)\; w(t) = V(L) \int_0^\infty p(\tau)\; w(t-\tau)d\tau \tag{2.5}$$

where we have defined $V(L) = C(L)^{-1} = \sum_{j=0}^{\infty} V_j L^j$, $V_0 = I$. Here $-V_j$ is the $n \times n$ matrix of coefficients on the jth lag in the vector autoregression for z. It follows directly upon writing out (2.5) that

$$a_t = \int_0^\infty f(\tau)w(t-\tau)d\tau \tag{2.6}$$

where[15]

$$f(\tau) = \sum_{j=0}^{\infty} V_j\, p(\tau - j) \tag{2.7}$$

It also follows from (2.6) and the identity for integer t, $C(L)\, a_t = P(D)\, w_t$, that

$$p(\tau) = \sum_{j=0}^{\infty} C_j\, f(\tau - j)\,. \tag{2.8}$$

Equations (2.6) and (2.7) show how the discrete time innovation a_t in general reflects all past values of the continuous time innovation $w(t)$.

Analyses of vector autoregressions often proceed by summarizing the shape of $C(L)$ in various ways, and attempting to interpret that shape. The innovation accounting methods of Sims, based on decomposition (7), are good examples of procedures that summarize the shape of $C(L)$. From the viewpoint of interpreting discrete time vector autoregressions in terms of the economic forces acting on individual agents, it would be desirable if the discrete time and continuous time moving-average representations were to match up in some simple and interpretable ways. In particular, the following two distinct but related features would be desirable. First, it would be desirable if the discrete time innovations a_t closely reflected the behavior of $w(s)$ near t. Probably the most desirable outcome would be if a_t could be expressed as

$$a_t = \int_0^1 f(\tau)\, w(t-\tau) d\tau\ , \tag{2.9}$$

so that in (2.6), $f(\tau) = 0$ for $\tau > 1$. In that case, a_t would be a weighted sum of the continuous time innovations over the unit forecast interval. It would be even more desirable if (2.9) were to hold with $f(\tau) = p(\tau)$, for then a_t would equal the one step ahead forecast error from the continuous time system. Second, assuming a smooth $p(\tau)$ function, it would be desirable if the discrete time moving-average coefficients $\{C_0, C_1, C_2, \ldots\}$ resemble a sampled version of the continuous time moving average kernel $\{p(\tau),\ \tau \geq 0\}$. This is desirable because the pattern of the C_j's would then faithfully reflect the response of the system to innovations in continuous time. We shall consider each of these *desiderata* in turn.

We first study conditions under which $f(\tau) = 0$ for $\tau > 1$. Consider the decomposition

$$\begin{aligned} a_t &= z(t) - \hat{E}\left[z(t) \mid w(t-s),\, s \geq 1\right] \\ &\quad + \hat{E}\left[z(t)|w(t-s),\, s \geq 1\right] - \hat{E}\left[z_t|z_{t-1}, \ldots\right] \\ &= \int_0^1 p(\tau)w(t-\tau)d\tau + \int_1^\infty p(\tau)w(t-\tau)d\tau \\ &\quad - \hat{E}\left[\int_1^\infty p(\tau)w(t-\tau)d\tau|z_{t-1}, \ldots\right] . \end{aligned}$$

This last equality implies that if (2.9) is to hold it must be the case that

$$\hat{E}[z(t)|w(t-s),\, s \geq 1] = \hat{E}[z_t|z_{t-1}, \ldots] , \tag{2.10}$$

which in turn implies that $p(\tau) = f(\tau)$ for $0 \leq \tau \leq 1$. The interpretation of requirement (2.10) is that the discrete time and continuous time forecasts of $z(t)$ over a unit time interval coincide.

When condition (2.9) is met, the link between $P(D)$ and $C(L)$ is particularly simple. Using $f(\tau) = 0$ for $\tau > 1$, equation (2.8) becomes

$$p(\tau) = C_j f(\tau - j) \quad \text{for} \quad j \leq \tau < j+1 . \tag{2.11}$$

Equation (2.11) implies that for the particular class of continuous time processes for which $f(\tau) = 0$ for $\tau > 1$, the continuous time moving-average coefficients are completely determined by the discrete time moving-average coefficients and the function $f(\tau)$ defined on the unit interval. The aliasing problem is manifested in this relationship because $f(\tau)$ cannot be inferred from discrete time data. In the absence of additional restrictions, all functions $f(\tau)$ that satisfy

$$\int_0^1 f(\tau)f(\tau)^T \, d\tau = W$$

are observationally equivalent. Relation (2.11) also implies that in general, without some more restrictions on $p(\tau)$, condition (2.9) does not place *any* restrictions on the discrete time moving-average coefficients.

However, in many (if not most) applications, it is usual to impose the additional requirement that the continuous time moving-average coefficients be a continuous function of τ.[16] This requirement together with (2.11) then imposes a very stringent restriction on the discrete time moving-average representation. In particular, (2.11) then implies that

$$C_j\, f(0) = C_{j-1}\, f(1) \tag{2.12}$$

where $f(\tau)$ is now a continuous function on the unit interval. When $w(t)$ and $z(t)$ have the same dimension ($m = n$) and $f(0)$ is nonsingular, relation (2.12) implies that

$$C_j = [f(1)f(0)^{-1}]^j$$

and

$$C(L) = [I - f(1)f(0)^{-1}L]^{-1} .$$

This implies that if (2.9) is to hold, the discrete time process must have a first order autoregressive representation. We have therefore established that condition (2.9) and the continuity requirement on $p(\tau)$ substantially restrict not only the admissible continuous time moving-average coefficients but the admissible discrete time moving-average coefficients as well.

Thus, with a continuous $p(\tau)$ function, in general, relation (2.9) does not hold. Instead, a_t given by (2.6) is a function of all current and past $w(t)$'s, a function whose nature can pose problems in several interrelated ways for interpreting a_t in terms of the continuous time noises $w(t)$ that are imagined to impinge on agents in the model. First, as in the discrete time case, the process $w(t)$ need not be fundamental for $z(t)$ in continuous time. Second, the matrix function $f(\tau)$ in (2.6) is not usually diagonal, so that each component of a_t in general is a function of all of the components of $w(t)$. This is a version of what Geweke (1978) has characterized as "contamination," which occurs in the context of the aggregation over time of several interrelated distributed lags. It is also related to the well-known phenomenon that aggregation over time generally leads to Granger-causality of discrete sampled y to x even when y fails to Granger-cause x in continuous time. Third, the matrix function $f(\tau)$ in (2.5) in general is nonzero for all values of $\tau > 0$, so that a_t in general depends on values of $w(t - \tau)$ in the remote past.

We now turn to our second *desideratum*, namely that the sequence $\{C_j\}_{j=0}^{\infty}$ resemble a sampled version of the function $p(\tau)$. For studying this matter, we set $m = n$, because we are interested in studying circumstances under which $\{C_j\}$ fails to reflect $p(\tau)$ even when the number of white noises n in a_t equals the number m in $w(t)$. We can represent most of the issues here with a univariate example, and so set $m = n = 1$ in most of our discussion. It is also convenient to study the case in which z_t has a rational spectral density in continuous time. Thus we assume that

$$\theta(D)z_t = \psi(D)\, w(t) \tag{2.13}$$

where z_t is a scalar stochastic process, and $\theta(s) = (s - \lambda_1)(s - \lambda_2) \ldots (s - \lambda_r)$, $\psi(s) = \psi_0 + \psi_1 s + \ldots + \psi_{r-1} s^{r-1}$. We assume that the real parts of $\lambda_1, \ldots, \lambda_r$, which are the zeroes of $\theta(s)$, are less than zero, but that the real parts of the zeroes of $\psi(s)$ are unrestricted. Only if the real parts of the zeroes of $\psi(s)$ are less than zero do current and past values of $z(t)$ and $w(t)$ span the same linear space. If any zeroes of $\psi(s)$ have real parts that exceed zero, then current and lagged $w(t)$ span a larger space than do current and lagged $z(t)$. The above equation can be expressed as

$$z_t = P(D)\, w(t) \tag{2.14}$$

where $P(D) = \psi(D)/\theta(D)$. A partial fraction representation of $P(D)$ is

$$P(D) = \sum_{j=1}^{r} \frac{\delta_j}{D - \lambda_j} \tag{2.15}$$

where

$$\delta_j = \lim_{s \to \lambda_j} P(s)\,(s - \lambda_j)\,. \tag{2.16}$$

We therefore have

$$p(\tau) = \sum_{j=1}^{r} \delta_j\, e^{\lambda_j \tau}\,. \tag{2.17}$$

Thus, the weighting function $p(\tau)$ in the continuous time moving-average representation is a sum of r exponentially decaying functions. Our object will now be to get an analogous expression to (2.17) for the discrete time coefficients B_k.

It is known that the discrete time process z_t implied by (2.13) is an rth order autoregressive, $(r - 1)$ order moving average process. Let this be $z_t = \frac{c(L)}{d(L)} a_t$ where $c(L) = \sum_{j=0}^{r-1} c_j L^j$, $d(L) = \sum_{j=0}^{r} d_j L^j$. To find this representation, we must use (2.4). A.W. Phillips (1959) and Hansen and Sargent (1983b) show that for the process (2.13), the term on the right side of (2.4) can be represented

$$\sum_{j=-\infty}^{\infty} P(i\omega + 2\pi i j)\, P(-i\omega - 2\pi i j) = \sum_{j=1}^{r} \left[\frac{w_j}{(1 - e^{\lambda_j} e^{-i\omega})} + \frac{w_j\, e^{\lambda_j}\, e^{+i\omega}}{(1 - e^{\lambda_j}\, e^{+i\omega})}\right]$$

where

$$w_j = \lim_{s \to \lambda_j} P(s)\, P(-s)\, (s - \lambda_j)\,.$$

Letting $z = e^{-i\omega}$, to find the required mixed moving-average autoregressive representation we must solve

$$\frac{c(z)c(z^{-1})}{d(z)d(z^{-1})} = \sum_{j=1}^{r} \left[\frac{w_j}{1 - e^{\lambda_j} z} + \frac{w_j e^{\lambda_j} z^{-1}}{1 - e^{\lambda_j} z^{-1}} \right] \tag{2.18}$$

subject to the condition that the zeroes of $c(z)$ and $d(z)$ all lie outside the unit circle. The term on the right side of (2.18) can be expressed as

$$\begin{aligned} &\frac{\sum_{j=1}^{r} w_j \prod_{k \neq j}^{r} (1 - \alpha_k z) \prod_{k=1}^{r} (1 - \alpha_k z^{-1})}{\prod_{j=1}^{r} (1 - \alpha_j z) \prod_{k=1}^{r} (1 - \alpha_k z^{-1})} + \\ &\frac{\sum_{j=1}^{r} w_j \alpha_j \prod_{k=1}^{r} (1 - \alpha_k z) \prod_{k \neq j}^{r} (1 - \alpha_k z^{-1}) z^{-1}}{\prod_{j=1}^{r} (1 - \alpha_j z) \prod_{k=1}^{r} (1 - \alpha_k z^{-1})} \end{aligned} \tag{2.19}$$

where $\alpha_j \equiv e^{\lambda_j}$. Note that $|\alpha_j| < 1$ by virtue of the assumption that real $(\lambda_j) < 0$. Thus, the denominator is already factored as required, so that

$$d(z) = \prod_{j=1}^{r} (1 - \alpha_j z)\,. \tag{2.20}$$

The numerator must be factored to find $c(z)$. Standard procedures to find the zeroes of scalar polynomials can be used to achieve this factorization, as described by Hansen and Sargent (1981a).

Thus we have that

$$z_t = \frac{c(L)}{d(L)}\, a_t \equiv C(L)\, a_t\,. \tag{2.21}$$

Proceeding in a similar fashion as we did for the continuous time moving-average representation, we can find a partial fraction representation for $C(L)$, namely

$$C(L) = \sum_{j=1}^{r} \frac{\gamma_j}{1 - \alpha_j L} \tag{2.22}$$

where

$$\gamma_j = \lim_{z \to \alpha_j^{-1}} C(z)\, (1 - \alpha_j z)\,. \tag{2.23}$$

Recalling that $\alpha_j = e^{\lambda_j}$, equation (2.22) implies that

$$C_k = \sum_{j=1}^{r} \gamma_j e^{\lambda_j k}. \tag{2.24}$$

Collecting and comparing the key results, we have that

$$p(\tau) = \sum_{j=1}^{r} \delta_j e^{\lambda_j \tau} \quad , \qquad \tau \in [0, \infty) \, . \tag{2.17}$$

$$C_k = \sum_{j=1}^{r} \gamma_j e^{\lambda_j k} \quad , \qquad k = 0, 1, 2, \ldots \tag{2.24}$$

Equations (2.17) and (2.24) imply that C_k will be (proportional to) a sampled version of $p(\tau)$ if and only if $\gamma_j / \delta_j = \gamma_1 / \delta_1$ for all $j = 2, \ldots, r$. It can be shown directly by using (2.17) and (2.24) in (2.7) and (2.8) that this condition will not be met for any $r \geq 2$. Thus, only if $z(t)$ is a first-order autoregressive process does C_k turn out to be a sampled version of $p(\tau)$.

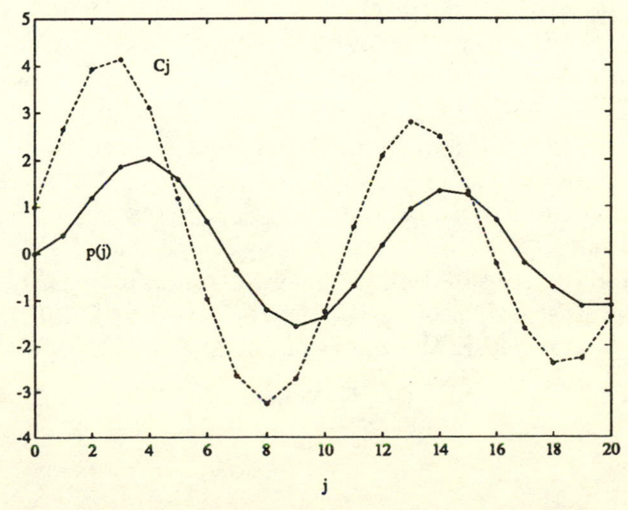

Figure 5. Continuous time ($p(j)$)and discrete time (C_j) moving average kernels.

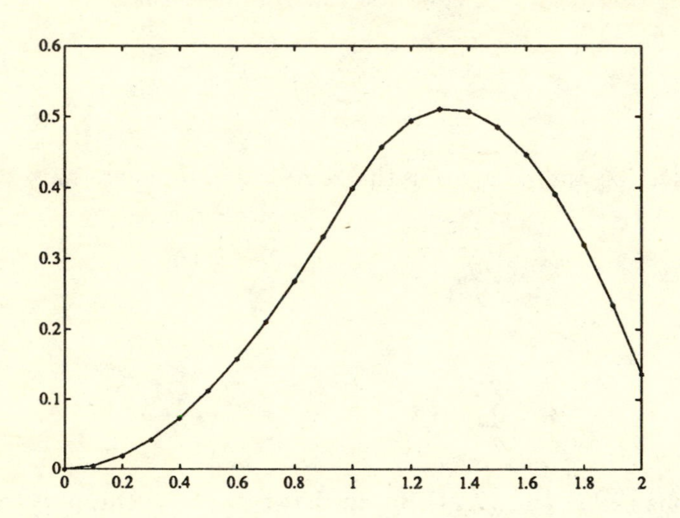

Figure 6. The function $f(s)$.

Table 1 and Figure 5 present a numerical example that illustrates the preceeding ideas. For the univariate process $(D^3 + .6D^2 + .4D + .2)\ z(t) = w(t)$, we have calculated $p(\tau)$, $f(\tau)$, $c(L)$, $d(L)$, $B(L) = c(L)/d(L)$, δ_j, γ_j for $j = 1, 2, 3$. In this example, we have that $\gamma_j/\gamma_1 \neq \delta_j/\delta_1$ for $j \geq 2$, so that the shapes of the moving averages in continuous and discrete time, $p(\tau)$ and C_k, respectively, are different. We plot C_k and $p(\tau)$ for integer values of τ in Figure 5. We also plot $f(\tau)$ in Figure 6. Notice that $f(\tau) \neq 0$ for some τ's greater than 1. In particular, notice that $f(\tau)$ is larger in absolute value over most of the interval [1, 2] than it is over the interval [0, 1]. The failure of $f(\tau)$ to be concentrated on [0, 1] and the failure of B_k to resemble a sampled version of $p(\tau)$ are both consequences of the fact that this is a third order autoregressive system in continuous time, rather than a first order one.

The preceding results and the example generalize readily to the case of a vector stochastic process z_t. Matrix versions of (2.17) and (2.24) hold, where the λ_j's are the zeroes of $\det \theta(s)$ and the δ_j's and the γ_j's are $(n \times n)$ matrices given by (2.16) and (2.23).

Table 1
An Example of Aggregation Over Time

$\psi(D) = 1$

$\theta(D) = .2 + .4D + .6D^2 + D^3$

λ_j (zeroes of $\theta(s)$): $-.5424$, $-.0288 \pm .6066$

δ_j in Partial Fraction Representation of $\psi(D)/\theta(D)$:

			δ_j/δ_1	
j	Real (δ_j)	Imaginary (δ_j)	Real	Imaginary
1	1.5831	0	1.00	0
2	$-.7915$	.6701	$-.50$	.423
3	$-.7915$	.6701	$-.50$	$-.423$

Zeroes of Spectral Factorization of Numerator Polynomial ($c(L)$):

	Real Part of Zero	Imaginary Part of Zero	Modulus
1	$-.0441$	0	.044
2	$-.4359$	0	.436

γ_j in Partial Fraction Representation of $C(L)$:

			γ_j/γ_1	
	Real (γ_j)	Imaginary (γ_j)	Real	Imaginary
1	1.7984	0	1.000	0
2	$-.3992$	2.0310	$-.222$	1.129
3	$-.3992$	-2.0310	$-.222$	1.129

Discrete Time Mixed Moving-Average, Autoregressive Representation:

$d(L) = 1 - 2.1779L + 1.8722L^2 - .5485L^3$

$c(L) = 1 + .4800L + .0192L^2$

Table 1 (continued)

τ	$f(\tau)$	$p(\tau)$	C_k
0	0	0	1.000000
.100	.004900	.004900	
.200	.019198	.010108	
.300	.042288	.042288	
.400	.073563	.073563	
.500	.112414	.112414	
.600	.158231	.153231	
.700	.210404	.210404	
.800	.268324	.268324	
.900	.331386	.331386	
1.000	.398987	.398987	2.657971
1.100	.457506	.470529	
1.200	.494395	.545421	
1.300	.510679	.623079	
1.400	.507397	.702926	
1.500	.485602	.784396	
1.600	.446360	.866935	
1.700	.390751	.949999	
1.800	.319860	1.033059	
1.900	.234786	1.115601	
2.000	.136629	1.197125	3.935901
3.000	−.073263	1.860267	4.144677
4.000	.032542	2.029242	3.116763
5.000	−.014212	1.593759	1.188521
6.000	.006197	.692895	−.972014
7.000	−.002701	−.361072	−2.631591
8.000	.001178	−1.208944	−3.259333
9.000	−.000513	−1.576723	−2.705194
10.000	.000224	−1.368770	−1.233866
11.000	−.000098	−.692635	.588609
12.000	.000043	.188765	2.107332
13.000	−.000019	.956663	2.810459
14.000	.000008	1.350008	2.498675
15.000	−.000004	1.252755	1.336741
16.000	.000002	.725963	−.224260
17.000	−.000001	−.020340	−1.619749
18.000	.000000	−.722582	−2.374213
19.000	−.000000	−1.131496	−2.261452
20.000	.000000	−1.124345	−1.369232

2a. Locally Unpredictable Processes and Linear Quadratic Models

The stochastic process $z(t)$ in Table 1 is mean square differentiable,[17] as evidenced by the fact that $p(0) = 0$. A stochastic process of the form (2.1) can be shown to be j times mean square differentiable if $p(0) = p'(0) = p''(0) = \ldots = p^{(j-1)}(0) = 0$ (see Sargent (1983) for a proof). Consequently, the process $(D^3 + .6D^2 + .4D + .2)z(t) = w(t)$ can be verified to be twice (but not three times) mean square differentiable. It is the smoothness and proximity to zero near $\tau = 0$ of $p(\tau)$ that makes it difficult for C_j to resemble a sampled version of $p(\tau)$, and that makes $a(t)$ a poor estimator of $\int_0^1 p(\tau)w(t-\tau)d\tau$.

Sims (1984) argued that there is a class of economic variables that are best modeled as failing to be mean square differentiable. For these processes, $p(0) \neq 0$. Processes of the form (2.1) in which $p(0) \neq 0$ are said to be *locally unpredictable* because if $p(0) \neq 0$, then

$$\lim_{\delta\to 0} \frac{E(x(t+\delta) - \hat{E}_t x(t+\delta))^2}{E(x(t+\delta) - x(t))^2} = 1\,. \tag{2.25}$$

Here $\hat{E}_t$ is the linear least squares projection operator, conditioned on $\{x(t-s), s \geq 0\}$. Now condition (2.25) can readily be shown to imply that

$$\lim_{\delta\to 0} \frac{E(x(t+\delta) - \hat{E}_t x(t+\delta))^2}{E(x(t+\delta) - \hat{E}(x(t+\delta)|x(t),\, x(t-\delta),\, x(t-2\delta),\, \ldots))^2} = 1 \tag{2.26}$$

In (2.26), $\hat{E}_t\, x(t+\delta)$ is the linear least squares projection of $x(t+\delta)$ conditioned on $(x(t-s), s \geq 0)$, while $\hat{E}(x(t+\delta)|x(t),\, x(t-\delta), \ldots)$ is the projection of $x(t+\delta)$ on the discrete time sample $x(t)$, $x(t-\delta), \ldots$. Condition (2.26) holds for any locally unpredictable process, and states that for small enough sampling interval δ, the δ-ahead projection error from the continuous time process is close in the mean square error sense to the δ-ahead projection error from the δ-discrete time data. Thus, when $p(0) \neq 0$, for small enough δ, the innovation a_t in the δ-counterpart to (2.21) is arbitrarily close to $\int_0^\delta p(s)w(t-s)ds$ in the mean square sense.

Now suppose that $z(t)$ is given by (2.1), with $p(0) = 0$, so that $z(t)$ is mean square differentiable. Following Sims (1984), suppose that the economist is interested in studying the expectational variable $x^*(t)$ given by

$$x^*(t) = \hat{E}\Big[\int_0^\infty e^{\rho s} z(t+s)ds|(z(t-\tau),\ t \geq 0)\Big] \tag{2.27}$$

where $\rho < 0$. Hansen and Sargent (1981d) showed that

$$\begin{aligned} x^*(t) &= \Big[\frac{-P(D)+P(-\rho)}{D+\rho}\Big]\, w(t) \equiv G(D)w(t) \\ &= \int_0^\infty g(s)w(t-s)ds\ , \end{aligned} \tag{2.28}$$

where $P(s) = \int_0^\infty e^{-\tau s} p(\tau)d\tau$ is the Laplace transform of $p(\tau)$. Now if $G(s)$ is the Laplace transform of $g(\tau)$, with support $[0, \infty)$, the initial value theorem for Laplace transforms states that

$$g(0) = \lim_{s\to\infty} s\, G(s)\ .$$

Using the initial value theorem together with (2.28), we find that

$$g(0) = \lim_{s\to\infty} s\Big[\frac{-P(s)+P(-\rho)}{s+\rho}\Big] = P(-\rho) \neq 0\ .$$

(We know that $P(-\rho) \neq 0$ because $P(s)$ is assumed to have no zeroes in the right half of the complex plane by the assumption that $p(\tau)$ is the kernel associated with a Wold representation for $z(t)$.) Therefore, even if $p(0) = 0$, $g(0) \neq 0$, so that the geometric expectational variable $x^*(t)$ fails to be mean square differentiable and therefore is locally unpredictable. For such expectational variables, (2.26) holds. Therefore, for such variables, for small enough sampling interval δ, the discrete time innovation $a(t)$ corresponding to (2.21) is close to $\int_0^\delta p(s)w(t-s)ds$ in the mean squared sense.

These results imply that for a variable $x^*(t)$ and sufficiently small sampling interval δ, the situation is not as bad as is depicted by the example in Table 1. As Sims has pointed out, there are theories of consumption and asset pricing which imply that consumption or asset prices behave like $x^*(t)$ and are governed by a version of (2.27). For example, with $x^*(t)$ being consumption and $z(t)$ income, (2.27) is a version of the permanent income theory. Alternatively, with $x^*(t)$ being a stock price and $z(t)$ being the dividend process, (2.27) is a simple version of an asset-pricing formula.

However, there is a wide class of generalized adjustment cost models discussed by Hansen and Sargent (1981a, 1981d) in which observable variables are such smoothed versions of $x^*(t)$ that they *are* mean square continuous. In adjustment cost models, decisions are driven by convolutions of $x^*(t)$, not by $x^*(t)$ alone. For example, the stochastic Euler equation for a typical quadratic adjustment cost problem is

$$(D-\rho)\,k(t) = E_t\,(\frac{1}{D+\rho})\,z(t)$$

where $\rho > 0$, or

$$(D-\rho)\,k(t) = x^*(t).$$

Here $k(t)$ is *capital.* The solution for capital is then

$$k(t) = \frac{1}{D-\rho}\,x^*(t)$$

or

$$k(t) = (\frac{1}{D-\rho})\,\Big[\frac{-P(D)+P(-\rho)}{D+\rho}\Big]\,w(t)$$

where $z(t) = \int_0^\infty \rho(s)\,w(t-s)$. Let

$$k(t) = \int_0^\infty h(\tau)w(t-\tau)d\tau$$

and

$$H(s) = \int_0^\infty e^{-\tau s}h(\tau)d\tau\ .$$

Then

$$H(s) = (\frac{1}{s-\rho})\ \Big(\frac{-P(s)+P(-\rho)}{s+\rho}\Big)\ .$$

Using the initial value theorem to calculate $h(0)$, we have

$$h(0) = \lim_{s\to\infty}\ sH(s) = 0\ .$$

Thus, $k(t)$ is mean square differentiable and so is locally predictable. (The convolution integration required to transform $x^*(t)$ to $k(t)$ *smooths* $k(t)$ relative to $x^*(t)$.)

More generally, the endogenous dynamics of adjustment cost models typically lead to mean square differentiable endogenous variables, provided that the agent is posited to be facing mean square differentiable forcing processes $(z(t))$. This means that for such models, the difficulties of interpretation that are illustrated in Table 1 cannot be eluded by appealing to an approximation based on the limit (2.26).

2b. Remedies in Continuous Time Analyses

The preceding problems of interpretation are results of estimating vector autoregressions while foregoing the imposition of any explicit economic theory in estimation. These problems can be completely overcome if a sufficiently restrictive and reliable dynamic model economy is available to be imposed during estimation. For example, Hansen and Sargent (1980b, 1981d) have described how the function $p(\tau)$ can be identified and estimated from observations on discrete time data in the context of a wide class of linear rational expectations models. The basic idea is that the rich body of cross-equation restrictions that characterize dynamic linear rational expectations models can be used to identify a unique continuous time model from discrete time data.

If an estimate of $p(\tau)$ is available, then by using only discrete time data on $\{z_t\}$, it is even possible to recover an estimate of the one-step prediction error that agents are making in continuous time. This is accomplished by treating the continuous time forecast error as a hidden variable whose covariances with the discrete time process $\{z_t\}$ are known. Thus, given estimates of $p(\tau)$, let us define the one-step ahead prediction error from continuous time data as $e_t^* = \int_0^1 p(\tau)\, w(t-\tau)\, d\tau$. Then it is straightforward to calculate the following second moments:

$$E(z_t z_{t-j}^T) = \int_0^\infty p(\tau+j)p(\tau)^T d\tau = \sum_{k=0}^{\infty} C_{k+j} W C_k^T,\ j \geq 0$$

$$E(e_t^*\, z_{t+j}^T) = \begin{cases} \int_0^1 p(\tau)p(\tau+j)^T d\tau & j \geq 0 \\ 0 & j < 0\,. \end{cases}$$

We can estimate the projection $\sum_{j=-m_1}^{m_2} D_j\, z_{t-j}$ in the projection equation

$$e_t^* = \sum_{j=-m_1}^{m_2} D_j\, z_{t-j} + u_t$$

where u_t is orthogonal to z_{t-j} for all $j = -m_1\, \ldots,\, m_2$. The D_j's can be computed from the normal equations

$$E\,(e_t^*\, z_{t+k}^T) = \sum_{j=m_1}^{m_2} D_j\, E\, z_{t-j}\, z_{t+k}^T,\ k = -m_2,\, \ldots,\, m_1\,.$$

These calculations could be of use if one's aim were truly to extract and to interpret estimates of the forecast errors made by agents. In continuous time versions of various models, such as those of Lucas (1973) or Barro (1977), agents' forecasting errors are an important source of impulses, so that it is of interest to have this method for characterizing their stochastic properties and estimating them.

3. Concluding Remarks

Subsequent chapters will treat aspects of the issues that we have studied in this chapter. The next chapter by Hansen, Roberds, and Sargent describes some discrete time tax and consumption models in which the history of innovations in a vector autoregression fails to equal the history of information possessed by agents. This poses problems in testing a key feature of the models, namely, a form of present value budget balance. For these particular models, the paper describes and implements a testing strategy that is an alternative to the *remedy in discrete time* described above.

Analysis of the issues raised in Section 2 on continuous time model is taken up and extended in Chapter 10 by Marcet, who relaxes our assumption that the continuous time spectral density is rational. This lets him study continuous time processes that have Wold representations with discontinuous moving average kernels. The remaining Chapters 7, 8, and 9 contain a variety of technical results that would be useful in implementing the *remedy in continuous time* described above.

Appendix

This appendix describes the recursive methods by which Figures 1–4 were computed.

We computed the objects described in Section 1 by mapping the model into the class of economies described by Hansen and Sargent (1990), and by using the computer programs that they describe. Hansen and Sargent (1990) describe a class of economies whose equilibrium allocations solve the social planning problem: choose stochastic processes $\{c_t,\ s_t,\ \ell_t,\ g_t,\ k_t,\ h_t\}_{t=0}^{\infty}$ to maximize

$$(A1) \qquad -\left(\frac{1}{2}\right) E_0 \sum_{t=0}^{\infty} \beta^t \left[(s_t - b_t) \cdot (s_t - b_t) + \ell_t^2\right]$$

subject to

$$
(A2) \quad
\begin{aligned}
\Phi_c c_t + \Phi_i i_t + \Phi_g g_t &= \Gamma k_{t-1} + d_t \\
g_t \cdot g_t &= \ell_t^2 \\
k_t &= \Delta_k k_{t-1} + \Theta_k i_t \\
h_t &= \Delta_h h_{t-1} + \Theta_h c_t \\
s_t &= \Lambda s_{t-1} + \Pi c_t \\
z_{t+1} &= A_{22} z_t + C_2 w_{t+1} \\
b_t &= U_b z_t \ , \ d_t = U_d z_t \ .
\end{aligned}
$$

In (A1), s_t is a vector of consumption service flows, b_t is a vector of preference shocks, ℓ_t is labor supply, c_t is a vector of consumption rates, i_t is a vector of investment rates, g_t is a vector of "intermediate goods," k_{t-1} is a vector of capital stocks, d_t is a vector of "endowment shocks," h_t is a vector of consumer durables, $\{w_{t+1}\}$ is a vector white noise with $E\, w_{t+1} w_{t+1}^T = I$, and z_t is an exogenous state vector of information. The planner maximizes A1 subject to A2 by choosing contingency plans for $\{c_t,\ i_t,\ \ell_t,\ g_t,\ k_t,\ h_t\}$ as functions of information known at t, namely $x_t^T \equiv [h_{t-1}^T,\ k_{t-1}^T,\ z_t^T]$.

There is a potential source of notational confusion because several sets of notations in $(A1)$–$(A2)$ were used to denote different objects in the model of Section 1. To avoid confusion, in this appendix we simply place a $(\bar{\ })$ above any variable in the model of Section 1 that might be confused with a similarly named variable in $(A1)$–$(A2)$.

We begin by eliminating p_t from equations (1.21) and (1.22) to obtain a single Euler equation in q_t:

$$
(A3) \quad
\begin{aligned}
E_t \Big\{ [h_d + g_d\, a(\beta L^{-1})\, a(L) + h_s + g_s(1 - \beta L^{-1})\,(1 - L)]\, q_t \Big\} \\
+ \bar{d}_t + \bar{s}_t = 0
\end{aligned}
$$

To obtain a version of the social planning problem $(A1)$–$(A2)$ whose quantities (and shadow prices) solve the model of Section 1, our strategy is to choose the objects in $(A2)$ so that the Euler equations for problem $(A1)$–$(A2)$ match up with $(A3)$.

Let $f_1,\ f_2,\ f_4,\ f_5,\ f_6,\ f_7$ be undetermined scalars which we shall eventually set in order to match $(A3)$ with the Euler equation for problem $(A1)$–$(A2)$. Then we propose the following settings for the objects in $(A2)$:

$$\begin{aligned} s_{1t} &= f_2\, q_t \qquad\qquad\qquad (q_t = c_t) \\ s_{2t} &= f_1\, [a(L)\, q_t] \\ &= f_1\, a_0\, q_t + f_1\, [a_1\ a_2\ a_3\ a_4] \begin{bmatrix} q_{t-1} \\ q_{t-2} \\ q_{t-3} \\ q_{t-4} \end{bmatrix} \\ b_{2t} &= 0 \\ b_{1t} &= f_4\, \bar{d}_t \end{aligned}$$

Thus, we set

$$\underbrace{\begin{bmatrix} s_{1t} \\ s_{2t} \end{bmatrix}}_{s_t} = \underbrace{f_1 \begin{bmatrix} 0 & 0 & 0 & 0 \\ a_1 & a_2 & a_3 & a_4 \end{bmatrix}}_{\Lambda} \underbrace{\begin{bmatrix} q_{t-1} \\ q_{t-2} \\ q_{t-3} \\ q_{t-4} \end{bmatrix}}_{h_{t-1}} + \underbrace{\begin{bmatrix} f_2 \\ f_1\, a_0 \end{bmatrix}}_{\Pi} \underbrace{q_t}_{c_t}$$

$$\underbrace{\begin{bmatrix} q_t \\ q_{t-1} \\ q_{t-2} \\ q_{t-3} \end{bmatrix}}_{h_t} = \underbrace{\begin{bmatrix} 0 & 0 & 0 & 0 \\ 1 & 0 & 0 & 0 \\ 0 & 1 & 0 & 0 \\ 0 & 0 & 1 & 0 \end{bmatrix}}_{\Delta_h} \underbrace{\begin{bmatrix} q_{t-1} \\ q_{t-2} \\ q_{t-3} \\ q_{t-4} \end{bmatrix}}_{h_{t-1}} + \underbrace{\begin{bmatrix} 1 \\ 0 \\ 0 \\ 0 \end{bmatrix}}_{\Theta_h} q_t$$

As for the technology, we specify

$$\begin{aligned} g_{1t} - f_5\, q_t &= f_6\, \bar{s}_t \\ g_{2t} - f_7\, i_t &= 0 \\ c_t &= k_t \\ k_t &= k_{t-1} + i_t \end{aligned}$$

To implement these specifications, we set

$$\underbrace{\begin{bmatrix} 1 \\ -f_5 \\ 0 \end{bmatrix}}_{\Phi_c} c_t + \underbrace{\begin{bmatrix} -1 \\ 0 \\ -f_7 \end{bmatrix}}_{\Phi_i} i_t$$

$$+ \underbrace{\begin{bmatrix} 0 & 0 \\ 1 & 0 \\ 0 & 1 \end{bmatrix}}_{\Phi_g} \begin{bmatrix} g_{1t} \\ g_{2t} \end{bmatrix} = \underbrace{\begin{bmatrix} 1 \\ 0 \\ 0 \end{bmatrix}}_{\Gamma} k_{t-1} + \begin{bmatrix} 0 \\ f_6\, \bar{s}_t \\ 0 \end{bmatrix}$$

where $k_t = 1 \cdot k_{t-1} + 1 \cdot i_t$, so that $\Delta_k = \Theta_k = 1$. We set $d_{2t} = f_6\, \bar{s}_t$.
With these settings, we have that

$$
\begin{aligned}
(s_t - b_t)\cdot(s_t - b_t) &= (f_2 q_t - f_4 \bar{d}_t)^2 + (f_1\, a(L)\, q_t)^2 \\
&= f_2^2 q_t^2 + f_4^2 \bar{d}_t^2 - 2 f_2 f_4 q_t \bar{d}_t + f_1^2\, [a(L)\, q_t]^2 \,.
\end{aligned}
$$

We also have that

$$
\begin{aligned}
g_t \cdot g_t &= (f_5 q_t + f_6 \bar{s}_t)^2 + f_7^2\, i_t^2 \\
&= f_5^2 q_t^2 + f_6^2 \bar{s}_t^2 + 2 f_5 f_6 q_t \bar{s}_t + f_7^2\, (k_t - k_{t-1})^2 \,.
\end{aligned}
$$

Thus, the social planning criterion $(A1)$ can be expressed as

$$
(A4) \qquad -.5\, E \sum_{t=0}^{\infty} \beta^t J_t
$$

where

$$
\begin{aligned}
J_t = \Big\{ & f_2^2\, q_t^2 + f_4^2\, \bar{d}_t^2 - 2 f_2 f_4 \bar{d}_t + f_1^2\, [a(L)\, q_t]^2 \\
& + f_5^2\, q_t^2 + f_6^2 \bar{s}_t^2 + 2 f_5 f_6\, q_t \bar{s}_t + f_7^2\, (k_t - k_{t-1})^2 \Big\}
\end{aligned}
$$

Applying techniques described in Sargent (1987b, ch. IX and XVI), the Euler equation for $(A4)$ is evidently

$$
(A5) \qquad
\begin{aligned}
E_t \Big\{ & f_2^2\, q_t - f_2 f_4 q_t \bar{d}_t + f_1^2\, [a(\beta L^{-1})\, a(L)]\, q_t \\
& + f_5^2\, q_t + f_5 f_6 \bar{s}_t + f_7^2\, (1 - \beta L^{-1})\, (1 - L)\, q_t \Big\} = 0 \,.
\end{aligned}
$$

To make $(A5)$ match $(A3)$, it suffices to use the following settings for the f_j's:

$$
\begin{aligned}
f_1 &= (g_d)^{1/2} & \qquad f_5 &= (h_s)^{1/2} \\
f_2 &= (h_d)^{1/2} & \qquad f_6 &= (h_s)^{-1/2} \\
f_4 &= -(h_d)^{-1/2} & \qquad f_7 &= (g_s)^{1/2} \,.
\end{aligned}
$$

As for the information specification, we set

$$
z_t^T = [w_{st},\, w_{st-1},\, w_{st-2},\, w_{st-3},\, w_{dt},\, w_{dt-1},\, w_{dt-2},\, w_{dt-3}]
$$

$$
A_{22} = \begin{bmatrix} C_4 & O_4 \\ O_4 & C_4 \end{bmatrix}, \quad C_2^T = \begin{bmatrix} 1 & 0 & 0 & 0 & 0 & 0 & 0 & 0 \\ 0 & 0 & 0 & 0 & 1 & 0 & 0 & 0 \end{bmatrix}
$$

where C_4 is the (4×4) companion matrix of $[0\ 0\ 0\ 0]$, and O_4 is the (4×4) zero matrix. We set

$$U_b = f_4^{-1} \begin{bmatrix} bs & O_{1\times 4} \\ O_{1\times 8} & \end{bmatrix}, \quad U_d = f_6^{-1} \begin{bmatrix} O_{1\times 8} & \\ O_{1\times 4} & bd \\ O_{1\times 8} & \end{bmatrix}$$

where

$$bs = [B_{s0},\ B_{s1},\ B_{s2},\ B_{s3}]$$
$$bd = [B_{d1},\ B_{d2},\ B_{d3},\ B_{d4}]$$

where the B_{sj} and B_{dj} are defined in (1.23). We set U_b and U_d as above, because we want to implement $b_{1t} = f_4\, \bar{d}_t$ and $d_{2t} = f_6\, \bar{s}_t$ or equivalently, $\bar{d}_t = f_4^{-1}\, b_{1t}$ and $\bar{s}_t = f_6^{-1}\, d_{2t}$.

A solution of the social planning problem is computed by mapping A1 – A2 into a discounted dynamic programming problem. The solution is represented in the form of the stochastic difference equation

$$x_{t+1} = A^o\, x_t + C\, w_{t+1}\ , \tag{A6}$$

where

$$x_t = \begin{bmatrix} h_{t-1} \\ k_{t-1} \\ z_t \end{bmatrix}.$$

Hansen-Sargent (1990) show how to decentralize the social planning problem via a competitive equilibrium. It turns out that the spot price of consumption in the general equilibrium model of Hansen-Sargent equals p_t of the model of Section 1.[18] It also turns out that p_t is simply a linear function of the state x_t, as is q_t. The marginal utility of consumption ($q_t = c_t$) for this model can be shown to be given by

$$-E_t\, \{h_d + g_d\, a(\beta L^{-1})\, a(L)\}\, q_t - \bar{d}_t\ ,$$

which equals p_t by virtue of equation (1.22). Hansen and Sargent (1990) compute the marginal utility of consumption in the form $M_c\, x_t$, where M_c is a matrix comformable to x_t. Thus, we compute the price for our model simply as $p_t = M_c\, x_t$. Similarly, the level of consumption, q_t, is represented as $q_t = S_c\, x_t$, where Hansen and Sargent (1990) describe how to compute S_c.

Hansen-Sargent's computations can be used to represent the solutions of the model in the state space form

$$\begin{aligned} x_{t+1} &= A^o\, x_t + C\, w_{t+1} \\ z_t &= G\, x_t \end{aligned} \tag{A7}$$

where $z_t = \begin{bmatrix} q_t \\ p_t \end{bmatrix}$ and where $G = \begin{bmatrix} S_c \\ M_c \end{bmatrix}$. Representation A7 is a state-space version of representation (1.13). We computed the impulse response function from (A7) to create Figure 1, which gives the impulse response of $S(L)\, z_t = R(L)\, \varepsilon_t$.

Using the Kalman filter, we obtained the following *innovations representation* associated with (A7):

$$(A8) \qquad \begin{aligned} \hat{x}_{t+1} &= A^o\, \hat{x}_t + K a_t \\ z_t &= G\, \hat{x}_t + a_t \end{aligned}$$

where $a_t = z_t - E\,[z_t \mid z_{t-1}, z_{t-2}, \ldots]$, $\hat{x}_t = E[x_t | z_{t-1}, z_{t-2}, \ldots]$, and K is the Kalman gain. Representation (A8) corresponds to the Wold (fundamental) moving average representation for z_t given by representation (1.17). The innovation a_t in (A8) equals $R_0^*\, \varepsilon_t^*$ of (1.17). We created Figures 2 and 3 by computing impulse response functions from (A8).

To create Figure 4, we used (A7)–(A8) as a coupled system. First, we *turned around* A8 to achieve the *whitener*

$$(A9) \qquad \begin{aligned} \hat{x}_{t+1} &= (A^o - KG)\, \hat{x}_t + K\, z_t \\ a_t &= \; -G\, \hat{x}_t + z_t \, . \end{aligned}$$

Then we created the coupled system formed by taking the *output* z_t of (A7) as the *input* of system (A9). We then used the coupled system to compute the impulse response functions of a_t (i.e., $R_t^*\, \varepsilon_t^*$) with respect to w_t (i.e., ε_t) that are reported in Figure 4.

Notes

1. This is, after all, the construction used in Wold's decomposition theorem.

2. Throughout this paper, we use $\hat{E}$ to denote the linear least squares projection operator.

3. Representations of the moving average in (1) are not in general unique once one relaxes the restriction in (1) that $A_0^{\Delta} = -I$, which in turn implies that $C_0^{\Delta} = I$. If this restriction is relaxed, then any representation generated by slipping a UU^T in between C_j^{Δ} and

$a(t-\Delta j)$ in (4), where U is a unitary matrix ($UU^T = I$), is also a fundamental moving-average representation. That is,

$$z(t) = \sum_{j=0}^{\infty} (C_j^{\Delta} U)\,(U^T a(t-\Delta j))$$

is also a fundamental moving-average representation, since $U^T\ a(t)$ spans the same linear space as $a(t)$. In terms of such a representation, the decomposition of prediction error covariance becomes

$$E(z(t) - \hat{E}_{t-j}z(t))\,(z(t) - \hat{E}_{t-j}z(t))^T$$
$$= \sum_{k=0}^{j-1} C_j^{\Delta}\ U\ V\ U^T\ C_j^{\Delta T},$$

which is altered by alternative choices of U. Sims' choice of orthogonalization order amounts to a choice of U.

4. An earlier version of this paper considered four classes of examples, the other two being nonlinearities and aggregation across agents. Due to length constraints, we decided to restrict this paper to the two classes of examples studied here.

5. Danny Quah has conveyed to us the viewpoint that implicit in the desire to match the $\{\varepsilon_t\}$ process of the economic model (1.8) with the $\{a_t\}$ process of the vector autoregression (1) must be a decision problem that concerns the data analyst. For example, on the basis of variance decompositions based on (7), the analyst might want to predict the consequences of *interventions* in the form of alterations in various diagonal elements of the innovation covariance matrix V, interpreting these alterations, e.g., as changes in the predictability by agents of various economic process, such as the money supply.

6. This assumption is made in the interest of providing the best possible chance that the processes $\{a_t\}$ and $\{\varepsilon_t\}$ match up. If ε_{1t} is a vector of dimension greater than n_1, then in general current and lagged values of $(\varepsilon_{1t}, \varepsilon_{2t})$ span a larger linear space than do current and lagged values of (y_t, x_{2t}).

7. See Rozanov (1967).

8. This inequality means that $E(R_0^* \varepsilon_t^*)(R_0^* \varepsilon_t^*)^T - E(R_0 \varepsilon_t)(R_0 \varepsilon_t)^T$ is a positive definite matrix.

9. The appendix maps the current model into a social planning problem that is a special case of the one studied by Hansen and Sargent (1990). By using Hansen and Sargent's interpretation of their setup as a general competitive equilibrium, it is possible to produce a general equilibrium interpretation of the model in the text.

10. It is interesting to note that although this system is one in which there are not strictly econometrically exogenous variables, or even any variables that are not Granger-caused by any others, its parameters are in principle identifiable. Identification is achieved through the cross-equation restrictions. Even when (w_{st}, w_{dt}) lie in the space spanned by the one-step ahead errors in predicting (q_t, p_t) from their own pasts, it is necessary to know the structural parameters of the model in order to deduce the former from the latter innovations.

11. The computational methods are described in the appendix. Briefly, we proceeded by mapping the problem into the general setup of Hansen and Sargent (1990), and using their computer programs.

12. By increasing the absolute values of the zeros of the polynomials $B_d(L)$ and $B_s(L)$ we were able to generate more "spectacular" examples of the phenomenon under discussion, in the sense that the discrepancy between the two covariance matrices in inequality (1.19) was even larger.

13. See Sims (1980) for a treatment of orthogonalization orders. Different "orthogonalization orders" in the sense of Sims amount to different triangular choices of the orthogonal matrix U that appears in footnote 3. If U^T is chosen to be upper triangular, then the first component of a_t corresponds to the first component of the new (basis) fundamental noise $U^T a_t$. On the other hand, if U^T is chosen to be lower triangular, the last component of a_t gets to go first in the Gram-Schmidt process that is used to create $U^T a_t$.

14. Practical methods for solving this equation for the case in which $P(s)$ is rational are discussed by Phillips (1959), Hansen and Sargent (1980b), and Christiano (1980).

15. An alternative derivation of (2.7) uses operational calculus. Setting $L = e^{-D}$, express (2.5) as $a_t = V(e^{-D})\ P(D)\ w(t) \equiv f(D)\ w(t)$. Here the function $f(\tau)$ is the inverse Fourier transform of $F(i\omega)$, which is defined by

$$F(i\omega) = C(e^{-i\omega})^{-1}\, P(i\omega)\ .$$

Equation (2.7) follows from the above equation by the convolution property of Fourier transforms.

16. For example, the function $p(\tau)$ will be continuous whenever $P(D)$ is rational, a common specification in applied work. The functions $p(\tau)$ and $f(\tau)$ are only defined up to an L^2 equivalence. Consequently, we can only impose continuity on one version of the continuous time moving average coefficients.

17. See Sargent (1983) for definitions of mean square continuity and mean square differentiability.

18. The spot price of consumption is p_t^t in the language of Hansen-Sargent.

5

Time Series Implications of Present Value Budget Balance and of Martingale Models of Consumption and Taxes

by Lars Peter HANSEN, William ROBERDS and Thomas J. SARGENT

1. Introduction

Let $\{(r_t, p_t)\}$ be a covariance stationary process where $r_t - p_t$ is the *net surplus* at time t, k_t a level of assets (or debts) carried into period t, and $(1 + \delta_t)$ the gross rate of return on those assets between t and $t+1$. Assume that there is a sequence of budget constraints

$$k_{t+1} = (1 + \delta_t)\, k_t + r_t - p_t, \quad t = 0, 1, \ldots \tag{1.1}$$

with k_0 given. This paper studies the observable implications of some models which impose a terminal condition on assets $\{k_t\}$ which has the effect of converting (1.1) into an intertemporal budget constraint, one that asserts that for each $t \geq 0$, k_t equals present value of current and future surpluses, discounted at rates generated by the process $\{\delta_t\}$.

We study the implications of imposing two alternative assumptions on δ_t. For most of the paper, we focus on the first assumption, which is that $\delta_t = \delta$ for all $t \geq 0$. In Section 7, we briefly study a second assumption, that $E_{t-1}\, \delta_t = \delta$ for all $t \geq 0$. We characterize the restrictions on the bivariate process $\{(r_t, p_t)\}$ under both assumptions and characterize the restrictions on the trivariate process $\{(k_{t+1}, r_t, p_t)\}$ under the second assumption on δ_t. Presumably, an econometrician who possesses data on $\{(r_t, p_t)\}$, but not on $\{k_{t+1}\}$, would want to use the restrictions imposed on the bivariate process to arrive at a judgment of whether the data are consistent with present value budget balance.

We are motivated to obtain these characterizations because of our interest in two types of models, each of which incorporates a version of the intertemporal budget constraint induced by (1.1) and the terminal condition on $\{k_t\}$. The first type of model consists only of the intertemporal budget constraint, and seeks to test whether observations on the joint process $\{(r_t, p_t)\}$ or on the joint process $\{(k_{t+1}, r_t, p_t)\}$ satisfy the budget constraint. In this literature, represented by contributions by Hamilton and Flavin (1986), Hakkio and Rush (1986), Shim (1984), and Sargent (1987b), r_t has been interpreted as government expenditures, p_t as taxes and k_t as government debt. This literature can be interpreted as getting at the question either of whether the terminal constraint is operative (e.g., Hamilton and Flavin 1986), or whether the series on government expenditures r_t and tax collections p_t are measured in the way that is required to make the present value constraint implied by (1.1) hold (e.g., Shim 1984).[1] This literature also seeks to determine what time invariance and finite dimensional state restrictions on the joint stochastic process $\{(p_t, r_t)\}$ and the interest rate process δ_t must be imposed in order to give observable content to the present value budget constraint.

Under the assumption that $\delta_t = \delta$ for all t, it was shown by Sargent (1987b) that the present value restriction induces the restriction on a *particular* moving-average representation for $\{(r_t, p_t)\}$ that for each innovation, the present value of the response of $\{(r_t - p_t)\}$ is zero. But even if the budget restriction is true, it is possible to specify many other moving-average representations for $\{(r_t, p_t)\}$ that violate this restriction. Indeed, as we remark below, any vector autoregressive representation for $\{(r_t, p_t)\}$ *must* correspond to a moving-average representation that violates this restriction. However, moving-average representations are not unique. It turns out that for *any* jointly covariance stationary process $\{(r_t, p_t)\}$, there exist moving-average representations that *do* satisfy the restriction. These results characterize the sense in which even with ample time-invariance imposed, the present value restriction by itself is observationally empty and are described in Section 2.

The second type of model which includes a version of our present value budget constraint is a linear quadratic version of Hall's (1978) martingale model of consumption. In that model, δ_t is constant over time, p_t is consumption, r_t is labor income or an endowment shock, and k_t is the level of household nonhuman assets or the capital stock. A linear version of Barro's (1979) model of tax smoothing is isomorphic to Hall's, with δ again constant, p_t being tax collections, r_t being government expenditures, and k_t being government debt at time t. Present value budget balance is among the restrictions imposed by Hall and Barro's model. In Section 3, we characterize what, if anything, the present value budget balance restriction adds to the martingale restriction. We show that the conjunction of the martingale hypothesis with the present value hypothesis gives added content to the latter. In effect, the martingale restriction allows the econometrician to pin down one component of the information set of private agents, isolating an innovation to which the present value budget restriction on moving average responses does apply.

Section 4 extends the results of Section 3 by considering a richer class of models that, for the consumption interpretation of the models, assume a type of nonseparable preferences for consumption goods. This class of models permits aggregate consumption to be broken into several components, each of which is of different durability in the sense that it gives rise to a different time profile of service flows. We show that even in this richer framework, present value budget balance continues to impose an additional restriction over and above the martingale restrictions imposed by the Euler equations associated with the optimum problem. Section 5 focuses on a consumption interpretation of the model, though presumably there is also a *tax smoothing* interpretation, with different components of consumption being reinterpreted, *a la* Barro, as different components of government revenues.

Section 5 implements the Section 4 tests for U.S. data on consumption and income. The tests turn up no evidence against the present value budget balance restriction. It would be useful to perform such tests for data on U.S. government expenditures and tax receipts, but we do not execute those tests here.

Section 6 briefly uses a version of ideas introduced by Sims (1972a) to study how issues of approximation bear on the interpretation of tests of our Section 4 restrictions. We show that even though those restrictions have content, they are very tenuous because of the existence of a sequence of false models that satisfy the restriction, and that approximates arbitrarily well a process that is known to violate the restrictions. This means that in practice rejections of the restrictions hinge on adopting sufficiently parsimonious specifications for the observable stochastic processes.

Section 7 focuses on the restrictions imposed by present value budget balance on the joint processes $\{(k_{t+1}, r_t, p_t)\}$ and $\{(r_t, p_t)\}$ under the assumption that δ_t is stochastic but satisfies $E_{t-1}\,\delta_t = \delta$ for all t. The specification $E_{t-1}\,\delta_t = \delta$ does not restrict $\{(r_t, p_t)\}$ but does result in an *exact linear rational expectations model* for the joint process $\{(k_{t+1}, r_t, p_t)\}$. We briefly compare these latter restrictions with those proposed by Hamilton and Flavin (1986). In the next paper in this volume (Chapter 6), Roberds uses data on the U.S. Federal budget to test present value budget balance by using an exact linear rational expectations model for the process $\{k_{t+1}, r_t - p_t\}$.

2. Implications of Present-Value Budget Balance

Let $s_t = r_t - p_t$. Depending in the particular model at hand, $\{s_t\}$ is a stochastic process either of receipts minus expenditures or of expenditures minus receipts. There is a sequence of budget constraints:

$$k_{t+1} = (1+\delta)k_t + s_t \qquad \text{for } t = 0, 1, \ldots \tag{2.1}$$

where k_0 is an initial condition, k_t is a measure of an asset or debt stock at time t, and $(1+\delta)$ is the gross rate of return between time t and time $t+1$. This return is assumed to be constant. In a permanent income model for consumption, let k_t be a consumer's assets at the beginning of period t, r_t exogenous labor income, and p_t consumption. In a model of the government budget, we let k_t be the stock of goverment debt at the beginning of t, r_t the level of government expenditures, and p_t the level of government tax collections. When the initial condition k_0 is observed by an econometrician, $\{k_t\}$ can be generated using (2.1). However, we assume that this initial level is not observed, so that $\{k_t\}$ is not observed by the econometrician.

Without constraining the process $\{k_t\}$, (2.1) is evidently not restrictive. In fact, we can just use (2.1) to define k_{t+1} recursively as a function of k_t and s_t for any initial condition k_0 and any process $\{s_t\}$. We are interested, however, in situations in which there is a terminal constraint imposed on asset holdings that, in effect, converts (2.1) into a discounted present-value budget constraint. For example, suppose that (2.1) holds for $t = 0, 1, \ldots T$ and that k_{T+1} is constrained to be zero. Then we can write

$$k_0 = -\sum_{t=0}^{T} \lambda^{t+1} s_t\,, \tag{2.2}$$

where $\lambda \equiv 1/(1+\delta)$. Taking limits as T goes to infinity gives

$$k_0 = -\sum_{t=0}^{\infty} \lambda^{t+1} s_t\,, \tag{2.3}$$

where the infinite series on the right side is assumed to be mean-square convergent. We view (2.3) as the infinite horizon counterpart to the terminal condition that k_{T+1} be zero.

Let L^2 be the space of all scalar stochastic processes $\{x_t\}$ such that

$$E\sum_{t=0}^{\infty} \lambda^t (x_t)^2 < \infty. \tag{2.4}$$

Throughout all of our analysis, we maintain that $\{s_t\}$ is in L^2. This restriction is sufficient for the right side of (2.3) to be a well-defined mean-square limit. It can accommodate growth in $\{s_t\}$ as long as the growth is dominated appropriately by the discount factor λ. For example, let μ be a common growth factor for $\{s_t\}$. Multiply both sides by μ^{-t}, which gives

$$\mu^{-t} k_{t+1} = [(1+\delta)/\mu]\mu^{-t+1} k_t + \mu^{-t} s_t \,. \tag{2.5}$$

Let variables with * superscripts be scaled by μ^{-t} to remove the effect of the geometric growth, and let δ^* be constructed so as to satisfy $(1+\delta^*) = [(1+\delta)/\mu]$. Then (2.4) can be expressed as

$$k^*_{t+1} = (1+\delta^*) k^*_t + s^*_t \,, \tag{2.6}$$

which is a version of (2.1). Our subsequent analysis can be thought of as applying to the * variables, although for notational simplicity we will omit the *'s.

Alternatively, we can accommodate other forms of stochastic growth that can be eliminated by taking appropriate *quasi-differences*. Let $\alpha(L)$ be a quasi-differencing filter with a finite order ℓ. Apply $\alpha(L)$ to both sides of equation (2.1). In this case, let * variables denote variables to which $\alpha(L)$ has been applied. Then

$$k^*_{t+1} = (1+\delta) k^*_t + s^*_t \qquad \text{for } t = \ell,\ \ell+1, \ldots \,. \tag{2.7}$$

For convenience, we shift the starting point back from ℓ to zero, and again we omit the *'s.

Replicating the analysis leading up to (2.3) for any initial period t gives

$$k_t = -\sum_{\tau=0}^{\infty} \lambda^{\tau+1} s_{t+\tau} \,. \tag{2.8}$$

where the sequence $\{k_t\}$ is in L^2. In the remainder of this section, we focus on the following question: given a process $\{y_t\}$ that includes $\{s_t\}$ as one of its components, under what set of circumstances can we construct a process $\{k_t\}$ that satisfies (2.8) and that is *predetermined* in the sense that k_t depends only on random variables realized at time $t-1$ and earlier.

Consider an environment in which there is a covariance stationary, n-dimensional vector martingale difference sequence $\{w_t\}$ used to generate information in the economy. The time t information set J_t is generated by $w_t, w_{t-1}, \ldots$ for each t.[2] We let H be a subspace of L^2 containing processes $\{x_t\}$ that are adapted to $\{J_t\}$ in the sense that x_t is in J_t for each t. The restriction that k_t be predetermined is formalized as the restriction that $\{k_{t+1}\}$ be in H.

For convenience, we suppose that $E(w_t w_t') = I$. The net surplus process $\{s_t\}$ (or some geometrically scaled or quasi-differenced version of this process) is the first component of an m-dimensional vector process $\{y_t\}$ that is assumed to be a time invariant, linear function of this martingale difference sequence:

Assumption A1: $y_t = C(L)w_t$ for $t = 0, 1, \ldots$ where $C(z) = \sum_{j=0}^{\infty} c_j z^j$ and $\sum_{j=0}^{\infty} |c_j|^2 < \infty$.

One possibility is that s_t is the only component of y_t, in which case m is one. More generally, y_t can contain other variables that are useful in forecasting future values of s_t and are observed by an econometrician. It is straightforward to show that when A1 is satisfied, $\{s_t\}$ is in H.

We shall use the idea that the stochastic process $\{y_t\}$ is *stochastically nonsingular* from the perspective of linear prediction theory. Informally, stochastic nonsingularity requires that no component of y_t can be expressed as a possibly infinite linear combination of current, past and future values of the other components. Formally, this requirement is stated in terms of the spectral density matrix of $\{y_t\}$. Assumption A1 implies that the following radial limit

$$C(\theta) \equiv \lim_{\eta \uparrow 1} C[\eta \exp(i\theta)] \tag{2.9}$$

exists for almost all θ in $(-\pi, \pi]$. Then the spectral density matrix for frequency θ is

$$S(\theta) \equiv C[\exp(-i\theta)]\, C[\exp(i\theta)]' . \tag{2.10}$$

In terms of the spectral density matrix, stochastic nonsingularity amounts to:

Assumption A2: $S(\theta)$ has rank m for almost all θ in $(-\pi, \pi]$.

We now deduce the restrictions on $\{y_t\}$ implied by the fact k_t as given by (2.8) is predetermined. Let the first row of $C(z)$ be denoted $\sigma(z)$. Then

$$s_t = \sigma(L)w_t \ . \tag{2.11}$$

It can be verified by substitution into (2.8) that a process $\{k_t\}$ that satisfies (2.8) is given by:[3]

$$\begin{aligned} &k_t = \kappa(L)w_t \\ &\text{where } \kappa(z) \equiv -\,\lambda\sigma(z)/(1-\lambda z^{-1}) = -\lambda z\sigma(z)/(z-\lambda) \ . \end{aligned} \tag{2.12}$$

However, in general, the function $\kappa(z)$ has a two-sided Laurent series expansion about $z = 0$ implying that k_t depends on current and future values of w_t. For this reason, k_t given by (2.12) may not be predetermined. The problem is that the function $\kappa(z)$ may have a pole (diverge to infinity) at $z = \lambda$. In the special case in which $\sigma(\lambda) = 0$, $\kappa(z)$ ceases to have a pole at λ and, in fact, $\kappa(z)$ has a one-sided power series expansion. In addition, $\kappa(0) = 0$ which guarantees that k_t depends only on past information. Thus, a necessary and sufficient condition for k_t to be predetermined relative to the sequence of information sets $\{J_t\}$ is that $\sigma(\lambda) = 0$, which we summarize as:

Restriction R1: $\sigma(\lambda) = 0$

We have thus established

Proposition 1: Suppose A1 is maintained and that economic agents have access to J_t at time t. Then $\{k_{t+1}\}$ given by (2.8) is in H if and only if R1 holds.

A version of this result was derived previously and interpreted by Sargent (1987b, pages 381–385).[4] Note that $\sigma(\lambda) = 0$ (or equivalently, $\sum_{j=0}^{\infty} \sigma_j \lambda^j = 0$) states the present value of the moving average coefficients of the surplus equals zero for each innovation in w_t.

In general, R1 rules out the possibility that the moving-average representation in A1 is a Wold representation of $\{y_t\}$. Recall that in a Wold representation the Hilbert spaces generated by $\{y_t, y_{t-1}, \ldots\}$ and $\{w_t, w_{t-1}, \ldots\}$ must be identical. When Assumption A2 is satisfied, a necessary and sufficient condition for these two spaces to be the same is given by the following condition:

Restriction R2: The rank of $C(z)$ is n for all $|z| < 1$.

Under A2, restrictions R1 and R2 cannot both be satisfied because R1 implies that $C(\lambda)$ can have at most rank $m - 1$. Hence for R1 and R2 to be compatible, n must be less than m. If n is less than m, A2 is violated and the process $\{y_t\}$ is stochastically singular. We summarize this finding in the following:

Proposition 2: Suppose A1 and A2 are maintained. Then R1 cannot hold for any C satisfying R2.

This finding has the practical implication that one ought not to test R1 by estimating versions of Wold representations as is done, for example, when estimating vector autoregressions. This is true even though we have assumed that $\{y_t\}$ is covariance stationary and so possesses a Wold representation. Restriction R1 applies to a moving-average representation that is necessarily distinct from the Wold representation. It follows that one cannot test R1 by examining directly the impulse response functions from a vector autoregression.

Is there any way that R1 can be tested? We now show that without additional restrictions the answer is no. Suppose that $\{y_t\}$ satisfies A1 and A2. We know from the Wold Decomposition Theorem for covariance stationary processes that there exists a moving-average representation that satisfies rank condition R2. Hence it is an implication of A1 and A2 that

$$y_t = C^*(L)w_t^* \tag{2.13}$$

where $C^*(z)$ satisfies assumption A2 and $\{w_t^*\}$ is an m-dimensional, covariance stationary white noise process with contemporaneous covariance matrix I.[5] In light of Proposition 2, C^* does not satisfy restriction R1.

To show that Restriction R1 is not testable, we demonstrate that it is always possible to build another moving-average representation, distinct from (2.13), that satisfies R1. Given $\{w_t^*\}$, first construct another m-dimensional serially uncorrelated process, say $\{w_t\}$, that depends on current and future values of $\{w_t^*\}$ and for which $Ew_t w_t' = I$. The process w_t satisfies:

$$w_t = D(L^{-1})' w_t^* \text{ where } D(z) = \sum_{j=0}^{\infty} d_j z^j, \quad \sum_{j=0}^{\infty} |d_j|^2 < \infty \tag{2.14}$$

and

$$D[\exp(i\theta)]\, D[\exp(-i\theta)]' = I \text{ for almost all } \theta. \tag{2.15}$$

Notice that $D[\exp(i\theta)]$ is a unitary matrix for almost all θ, and

$$w_t^* = D(L)w_t \tag{2.16}$$

so that w_t^* depends only on current and past values of w_t. Consequently, the Hilbert space generated by w_t, w_{t-1}, ... is no smaller than the space generated by w_t^*, w_{t-1}^*, ... and in fact is often strictly larger. We can represent y_t in terms of w_t via:

$$y_t = C(L)w_t \quad \text{where} \quad C(z) = C^*(z)D(z)\,. \tag{2.17}$$

In light of (2.17), we can show that Restriction R1 is not testable by establishing the existence of a function $D(z)$ satisfying (2.15) such that the first row of $C^*(\lambda)D(\lambda)$ is zero. We now propose two alternative ways in which this can be accomplished. One possibility is to construct $D(z)$ so that $D(\lambda)$ is a matrix of zeros. An example of such a $D(z)$ is

$$D(z) = [(z-\lambda)/(1-\lambda z)]\, I\,. \tag{2.18}$$

This choice of $D(z)$ satisfies (2.15) because

$$\frac{[\exp(i\theta)-\lambda]\,[\exp(-i\theta)-\lambda]}{[1-\lambda\exp(i\theta)]\,[1-\lambda\exp(-i\theta)]} = 1 \quad \text{for all } \theta. \tag{2.19}$$

A second possibility is to form an orthogonal matrix Q, the first column of which is a vector that is proportional to the first row of $C^*(\lambda)$ and has norm one. The remaining columns of Q are a set of $m-1$ orthonormal vectors that are orthogonal to the first row of $C^*(\lambda)$. Therefore all entries in the first row of $C^*(\lambda)Q$ are zeroes except for the first entry. The matrix Q is then used to build $D(z)$ as follows:

$$D(z) = Q \begin{bmatrix} (z-\lambda)/(1-\lambda z) & 0 \\ 0 & I \end{bmatrix} \tag{2.20}$$

where the matrix I has $m-1$ rows and columns. Using (2.19), it is straightforward to show that $D(z)$ satisfies (2.15). Notice that the first row and column of $C^*(\lambda)D(\lambda)$ contain all zeroes.

An equivalent way to construct $C(z)$ is as follows. Form the process $\{k_t\}$ using equation (2.8) and a composite process $\{x_t\}$ where $x_t' \equiv [y_t', k_t]$. This composite process is stochastically singular because k_t is an infinite linear combination of current and future values of s_t. Nevertheless, it possess a Wold moving-average representation in terms of an m-dimensional vector white noise $\{w_t^+\}$:

$$y_t = C^+(L)w_t^+\,, \quad k_t = \kappa^+(L)w_t^+\,, \tag{2.21}$$

where linear combinations of x_t, $x_{t-1}, \ldots$ generate the same Hilbert space as linear combinations of w_t^+, w_{t-1}^+, ... Let $\sigma^+(z)$ be the first row of $C^+(z)$. Since k_t satisfies (2.8),

$$\kappa^+(z) = -z\lambda\sigma^+(z)/(z-\lambda) . \tag{2.22}$$

Note that k_t depends only on current and past values of w_t^+ because (2.21) is a Wold representation for $\{x_t\}$. Therefore, $\sigma^+(z)$ must satisfy Restriction R1. In Appendix A we show that $C^+(z) = C^*(z)D(z)$ where $D(z)$ is given by (2.20) can be used in (2.21) in forming a Wold representation for $\{x_t\}$.

Summarizing these results we have established:

Proposition 3: Suppose A1 and A2 are maintained. It is always possible to find a moving-average representation $y_t = C^+(L)w_t^+$ that satisfies R1.[6]

Thus, Proposition 3 shows that R1 is not testable without additional restrictions to aid in the identification of w_t. Proposition 2 demonstrates that one commonly used device for identifying w_t is inappropriate when $\{y_t\}$ has full rank. In the next two sections we consider alternative ways to identify components of moving-average representations that can be used in testing R1.

3. The Martingale Model

In this section we impose considerably more structure on the problem. First, we decompose the surplus process $\{s_t\}$ into two components, *payouts* and *receipts*. The receipt process is specified exogeneously, but the payout process is modeled as the optimal decision process from a quadratic optimization problem subject to constraint (2.3).[7] The objective function for this optimization problem is designed to imply a martingale model for the payout process. This leads to a version of Hall's (1978) model of consumption or Barro's (1979) model of taxation. We show that beyond the martingale characterization for payouts, present-value budget balance delivers an additional restriction that is testable so long as the discount factor is known *a priori*.

Let $\gamma(L)$ be a stationarity inducing transformation for a receipt process $\{r_t\}$. In our analysis we take $\gamma(z)$ to be either 1 or $1-z$ in cases in which the stochastic process for receipts has a unit root. Other specifications of $\gamma(z)$ can be explored by mimicking the analysis in this section.

Assumption A3: $\gamma(L)r_t = \rho(L)w_t$ where $\rho(z) = \sum_{j=0}^{\infty} \rho_j z^j$, $\sum_{j=0}^{\infty} |\rho_j|^2 < \infty$.

We let $\{p_t\}$ denote the payout process. The net surplus process $\{s_t\}$ of Section 2 is given by

$$s_t = r_t - p_t. \tag{3.1}$$

The payout process $\{p_t\}$ is determined as the solution to an optimization problem with the objective being to maximize:

$$-E\left[\sum_{t=0}^{\infty} \lambda^t \, (p_t - b)^2\right] \tag{3.2}$$

where $b > 0$ subject to constraint (2.1) with k_0 given. The processes $\{p_t\}$ and $\{k_{t+1}\}$ are restricted to be in H. The solution to this problem is described in Sargent (1987b) and Hansen (1987). The optimal decision processes for p_t and k_{t+1} satisfy:

$$\begin{aligned} p_t &= \delta k_t + (1-\lambda)E\left(\sum_{j=0}^{\infty} \lambda^j r_{t+j} \mid J_t\right) \\ k_{t+1} &= k_t + r_t - (1-\lambda)\, E\left(\sum_{j=0}^{\infty} \lambda^j r_{t+j} \mid J_t\right) \end{aligned} \tag{3.3}$$

It can be verified that constraint (2.1) is satisfied adding the two equations in (3.3) together and rearranging terms.

To deduce implications for $\{p_t\}$, take the first equation in (3.3) at time $t+1$, subtract the same equation at time t and then substitute for $k_{t+1} - k_t$ from the second equation. This yields the following result of Flavin (1981):

(3.4)

$$p_{t+1} - p_t = (1-\lambda)\, E\left(\sum_{j=0}^{\infty} \lambda^j r_{t+j+1} \,|J_{t+1}\right) - (1-\lambda)\, E\left(\sum_{j=0}^{\infty} \lambda^j r_{t+j+1} \mid J_t\right).$$

Using a formula reported in Hansen and Sargent (1980a, 1981b), it follows that

$$p_t - p_{t-1} = [(1-\lambda)\rho(\lambda)/\gamma(\lambda)]w_t \quad t = 1,\ 2,\ \ldots \tag{3.5}$$

We now investigate implications of (3.5) for the process $\{s_t\}$. Stationarity in $\{s_t\}$ is induced by taking first-differences:

$$\begin{aligned} s_t - s_{t-1} &= r_t - r_{t-1} - p_t + p_{t-1} \\ &= \sigma(L)w_t \end{aligned} \tag{3.6}$$

where

(3.7) $$\sigma(z) \equiv (1-z)\rho(z)/\gamma(z) - (1-\lambda)\rho(\lambda)/\gamma(\lambda).$$

To relate the present analysis to the analysis in Section 2, we assume that $\{y_t\}$ contains at least two components ($m \geq 2$), with the first component being $(1-L)s_t$ and the second component being $\gamma(L)r_t$. Notice that $(1-z)/\gamma(z)$ is either $(1-z)$ or 1. Evaluating $\sigma(z)$ at $z = \lambda$, it is evident that $\sigma(z)$ satisfies R1. Finally, consider the special case in which $\rho(\lambda)$ is zero. In this case $p_t - p_{t-1} = 0$ which implies that $\{y_t\}$ is stochastically singular. Therefore, this case is eliminated from consideration by A3.

More generally, we consider a payout process that satisfies:

(3.8) $$p_t - p_{t-1} = \pi w_t$$

for some $\pi \neq 0$. Then $\{p_t\}$ is a martingale adapted to $\{J_t\}$. In this case, $\sigma(z)$ satisfies:

Restriction R3: $\sigma(z) = (1-z)\rho(z)/\gamma(z) - \pi$ for some row vector π.

Restriction R3 is the focus of Hall (1978) and Flavin (1981). Note that R1 and R3 imply (3.8) for $\pi = (1-\lambda)\rho(\lambda)/\gamma(\lambda)$. The question of interest in this section is whether R1 imposes any additional restrictions once R3 is satisfied. In other words, is the restriction

(3.9) $$\pi = (1-\lambda)\rho(\lambda)/\gamma(\lambda)$$

testable?

To answer this question, we construct an orthogonal matrix Q as follows. Since A2 is satisfied, π cannot be zero. The first column of Q is $\pi'/|\pi|$ and the remaining columns contain n-1 orthonormal vectors that are orthogonal to π'. Then πQ is row vector with $|\pi|$ in the first position and zeros elsewhere. Define

(3.10)
$$\sigma^{+}(z) \equiv \sigma(z)Q\,,\ \rho^{+}(z) \equiv \rho(z)Q\,,\ \pi^{+} \equiv \pi Q\,,\ \text{and}\ \ w_t^{+} \equiv Q' w_t\,.$$

The Q transformation is introduced so that the first entry of w_t^{+} is proportional to πw_t. Hence $p_t - p_{t-1}$ depends only on the first entry of w_t^{+}. Partition $\sigma^{+}(z) = [\sigma_1^{+}(z), \sigma_2^{+}(z)]$, $\rho^{+}(z) = [\rho_1^{+}(z), \rho_2^{+}(z)]$, $w_t^{+\prime} = [w_{1t}^{+}, w_{2t}^{+\prime}]$ where $\sigma_1^{+}(z)$, $\rho_1^{+}(z)$ and w_{1t}^{+} each have one entry. Write

(3.11)
$$\begin{aligned} s_t - s_{t-1} &= \sigma_1^{+}(L)w_{1t}^{+} + \sigma_2^{+}(L)w_{2t}^{+} \\ \gamma(L)r_t &= \rho_1^{+}(L)w_{1t}^{+} + \rho_2^{+}(L)w_{2t}^{+} \\ p_t - p_{t-1} &= |\pi| w_{1t}^{+}\ . \end{aligned}$$

An equivalent representation of (3.9) is

$$(1-\lambda)\rho_1^+(\lambda)/\gamma(\lambda) = |\pi| \tag{3.12}$$

and

$$\rho_2^+(\lambda) = 0. \tag{3.13}$$

First, we investigate whether (3.12) is testable. We use (3.11) to compute the regression of $\gamma(L)r_t$ onto current and past values of $p_t - p_{t-1}$ or equivalently onto current and past values of w_{1t}^+. These regressions are:

$$\begin{aligned}\gamma(L)r_t &= \rho_1^+(L)w_{1t}^+ + e_t \\ &= \beta(L)\,(p_t - p_{t-1}) + e_t\end{aligned} \tag{3.14}$$

where $\beta(z) \equiv \rho_1^+(z)/|\pi|$ and the regression error e_t is given by

$$e_t = \rho_2^+(L)w_{2t}^+. \tag{3.15}$$

The function $\beta(z)$ is identifiable for $|z| < 1$.[8] Consequently, (3.12) is a testable restriction given knowledge of the discount factor λ.[9]

Restriction R4: $(1-\lambda)\beta(\lambda)/\gamma(\lambda) = 1$.

Proposition 4: Suppose A1–A3 are maintained. If R1 and R3 are satisfied for $\pi \neq 0$, then R4 is also satisfied.

In Section 2 we showed that, by itself, present-value budget balance imposes no testable restrictions on $\{y_t\}$, essentially because pertinent innovations to agent's information sets could not be identified. How then does it occur that, in conjunction with the martingale restriction (R3), the present-value budget balance restriction acquires the content summarized in R4? This content emerges because the martingale restriction allows us to deduce one component of the new information that arrives at time t to economic agents, namely, $p_t - p_{t-1}$. The present-value budget balance restriction then applies to this component.[10] In the analysis in Section 2, none of the components of the new information was identified. As a consequence, the derived restrictions were not testable.

The presence of an additional implication implied by present-value budget balance also has been noted by West (1988).[11] West derived a variance inequality that is robust to the misspecification of the information set of economic agents used to forecast future values of r_t.

Restriction R4 shares this robustness and in fact implies West's variance inequality. The converse is not true, however. West's variance inequality does not imply R4 so that, in principle, R4 can be used to construct a statistical test with additional power.[12]

Finally, we investigate whether R4 exhausts the testable implications of R1 when $\{p_t\}$ is a martingale (i.e., when R3 is satisfied). This amounts to assessing the empirical content of (3.13) $[\rho_2^+(\lambda) = 0]$. Relation (3.13), however, is not testable for the same reasons discussed in Section 2 that $\sigma(\lambda) = 0$ is not testable. To see this, form an m-1 dimensional vector $\hat{y}_t$ by taking the least squares regression error of y_t on the Hilbert space generated by current and past values of $p_t - p_{t-1}$. Since the first entry of y_t is a linear combination of $r_t - r_{t-1}$ and $p_t - p_{t-1}$, its forecast error is a linear combination of current and past values of $\gamma(L)r_t$. Hence the m-dimensional vector stochastic process of regression errors is stochastically singular. We simply eliminate the first entry when forming $\hat{y}_t$ which means that e_t is now the first entry of $\hat{y}_t$. The analysis in Section 2 applies to the $(m-1)$-dimensional process $\{\hat{y}_t\}$ with e_t playing the role of s_t. We thus have:

Proposition 5: Suppose A1–A3 are maintained. If R3 and R4 are satisfied, then it is always possible to find a moving-average representation $y_t = \hat{C}(L)\hat{w}_t$ that satisfies R1.

Proposition 5 offers a word of caution for statistical tests of the restrictions implied by R1 and R3. For a given parameterization of ρ^+, one might consider testing whether $\rho_2^+(\lambda) = 0$. Holding fixed the particular finite-dimensional parameterization of ρ_2^+, this restriction can be tested using, say, a likelihood ratio test. The message from Proposition 5 is that such a test is not particularly interesting because there will always exist an alternative, observationally equivalent parameterization of ρ_2^+ that satisfies the restriction by construction. Restriction R4, however, is not subject to this same criticism.

4. Nonseparable Preferences

A potentially important defect of the model used in Section 3 is that preferences for the payout are assumed to be time separable. While this may not be a bad model when the payout consists only of consumption goods that are nondurable, there is no natural measure of receipts (or say labor income) that is matched to nondurable consumption in the manner assumed in Section 3. In this section we extend the martingale model described in Section 3 to allow for the possibility that the payout can be divided into a vector of consumption goods, and that preferences for these consumption goods are not necessarily separable either across goods or over time. We then show that the present-value budget balance restriction R1 still has a testable implication.

In their analyses of permanent income models, Campbell (1987) and West (1988) assume, in effect, that nondurable consumption is a fixed fraction of total consumption. We adopt a different approach in which the total payout is divided and invested in one of ℓ possible ways. For instance, the alternative investments might entail expenditures on alternative consumption goods such as durables, nondurables, and services. Hence

$$p_t = \mu' c_t \tag{4.1}$$

where μ is an ℓ-dimensional vector of positive numbers. Bernanke (1985) used this strategy in modeling simultaneously two consumption goods classifications, durables and nondurables.

Following Telser and Graves (1972), Hansen (1987) and Eichenbaum and Hansen (1990), we model each of these investments as generating an intertemporal bundle of ℓ different services. The bundling is a device for modeling intertemporal nonseparabilities in preferences, such as consumption durability and habit persistence, and nonseparabilities across the different goods classifications. More precisely, a vector of consumption goods c_t generates a corresponding service vector $b_\tau c_t$ at time $t + \tau$ for $\tau \geq 0$ where:

Assumption A4: $B(z) \equiv \sum_{\tau=0}^{\infty} b_\tau z^\tau$ is continuous and nonsingular on the domain $\{|z| \leq 1\}$.[13]

At time t the total quantity of consumption services is given by

$$h_t \equiv \sum_{\tau=0}^{\infty} b_\tau c_{t-\tau} = B(L)c_t \tag{4.2}$$

where investments prior to time zero are taken as initial conditions. This specification accommodates the linear-quadratic durable goods consumption models of Mankiw (1982) and Bernanke (1985) as special cases, as well as linear-quadratic versions of the habit persistence model of Ryder and Heal (1973).

The preferences used to induce the optimal payout process are specified as follows in terms of the ℓ-dimensional service process:

$$-(1/2)E \sum_{t=0}^{\infty} \lambda^t (h_t - g)' (h_t - g) \; . \tag{4.3}$$

Here g is an ℓ-dimensional vector of satiation points. As in the Sections 2 and 3, we impose (2.1) and require $\{k_{t+1}\}$ to be in H. In addition, $\{p_t\}$ and the entries of the processes $\{c_t\}$ and $\{h_t\}$ are restricted to be in H. Finally, the receipt process $\{r_t\}$ is assumed to satisfy A3.

We solve this optimization problem using the method of Lagrange multipliers. Let mk_t be the Lagrange multiplier associated with (2.1), mp_t be the multiplier associated with (4.1) and mh_t the multiplier associated with (4.2) at time t. The multiplier processes $\{mk_t\}$, $\{mp_t\}$, and the components of the process $\{mh_t\}$ are restricted to be in H. The Lagrangean is

(4.4)
$$\mathcal{L} = E\sum_{t=0}^{\infty} \lambda^t \Big\{-(1/2)(h_t - g)\cdot(h_t - g) + mp_t(\mu' c_t - p_t) + mh_t'[h_t - B(L)c_t] + mk_t[-k_{t+1} + (1+\delta)k_t + r_t - p_t]\Big\}.$$

The first-order conditions for p_t, k_t, c_t and h_t are:

$$-mp_t + mk_t = 0 \tag{4.5}$$
$$-mk_t + E(mk_{t+1} \mid J_t) = 0 \tag{4.6}$$
$$-E[B(\lambda L^{-1})' mh_t \mid J_t] + \mu mp_t = 0 \tag{4.7}$$
$$mh_t - (g - h_t) = 0\,. \tag{4.8}$$

The multiplier mk_t can be interpreted as the shadow valuation of the capital stock at the end of time t, while the multiplier mp_t is the (indirect) marginal utility of the payout at time t. First-order condition (4.5) indicates that these two multipliers should be equal. First-order condition (4.6) restricts the shadow valuation process for the capital stock to be a martingale. In light of (4.5) and (4.6), the marginal utility process for the total payout should be a martingale as in the model investigated in Section 3.

The multiplier mh_t is the time t marginal utility vector for services. First-order condition (4.7) relates the current (indirect) marginal utility for the payout to the current and expected future values of the marginal utility vector for services. This link reflects the technology for converting a payout today into services in current and future time periods.

We now use the connection between the marginal utilities to deduce the optimal process for $\{c_t\}$. First solve (4.7) for mh_t as a function of current and expected future values of mp_t. In light of A4, $B(\lambda L^{-1})$ has a one-sided forward inverse. Consequently,

$$mh_t = E\left\{[B(\lambda L^{-1})^{-1'}(\mu mp_t) \mid J_t]\right\}. \tag{4.9}$$

Since $\{mp_t\}$ is a martingale,

$$mh_t = B(\lambda)^{-1'}\mu mp_t \tag{4.10}$$

implying that that the vector of marginal utilities for services is also a martingale.[14] Substituting for mh_t from (4.8) and for h_t from (4.2) gives

$$B(L)c_t + B(\lambda)^{-1'}\mu mp_t - g = 0\,. \tag{4.11}$$

We compute c_t, k_{t+1}, and mp_t by solving the three equation system:

$$\begin{aligned} &B(L)c_t + B(\lambda)^{-1'}\mu mp_t - g = 0 \\ &E(mp_{t+1}|J_t) - mp_t = 0 \\ &k_{t+1} - (1+\delta)k_t - r_t + \mu' c_t = 0\,. \end{aligned} \tag{4.12}$$

Taking a linear combination of these three equations,

$$\begin{aligned} &(1-\lambda)\mu' B(\lambda)^{-1}[B(L)c_t + B(\lambda)^{-1'}\mu mp_t - g] \\ &+ \lambda[\mu' B(\lambda)^{-1} B(\lambda)^{-1'}\mu]\, E[(-L^{-1}+1)mp_t|J_t] \\ &- (1-\lambda)\,[(1-\lambda^{-1}L)k_{t+1} - r_t + \mu' c_t] = 0\,. \end{aligned} \tag{4.13}$$

Rearranging terms gives

$$\begin{aligned} &(1-\lambda)\mu'[B(\lambda)^{-1}B(L) - I]c_t \\ &+ [\mu' B(\lambda)^{-1}B(\lambda)^{-1'}\mu]\, E[(1-\lambda L^{-1})mp_t|J_t] \\ &- (1-\lambda)\,(1-\lambda^{-1}L)\,k_{t+1} + (1-\lambda)r_t - (1-\lambda)\mu' B(\lambda)^{-1} g = 0\,. \end{aligned} \tag{4.14}$$

Note that $(1-\lambda)\mu'[B(\lambda)^{-1}B(z) - I]$ can be factored

$$(1-\lambda)\mu'[B(\lambda)^{-1}B(z) - I] = (z-\lambda)\chi(z) \tag{4.15}$$

where $\chi(z)$ has a power series expansion for $|z| < 1$, because $B(\lambda)^{-1}B(z) - I$ is a matrix of zeros when $z = \lambda$. Applying the forward operator $(1-\lambda L^{-1})^{-1}$ to (4.14) and taking expectations conditioned on J_t gives

$$\begin{aligned} &\chi(L)c_{t-1} + [\mu' B(\lambda)^{-1}B(\lambda)^{-1'}\mu]mp_t + \delta k_t \\ &+ (1-\lambda)E\,(\sum_{j=0}^{\infty}\lambda^j r_{t+j} \mid J_t) - \mu' B(\lambda)^{-1} g = 0\,. \end{aligned} \tag{4.16}$$

Since A4 is satisfied, $[\mu' B(\lambda)^{-1} B(\lambda)^{-1'} \mu]$ is strictly positive. Solving for mp_t,

$$\begin{aligned} mp_t = & - \epsilon\chi(L)c_{t-1} \\ & - \epsilon\Big[\delta k_t + (1-\lambda)E\Big(\sum_{j=0}^{\infty} \lambda^j r_{t+j} | J_t\Big)\Big] \\ & + \epsilon\mu' B(\lambda)^{-1} g \end{aligned} \tag{4.17}$$

where

$$\epsilon \equiv 1/[\mu' B(\lambda)^{-1} B(\lambda)^{-1'} \mu] \,. \tag{4.18}$$

We can obtain recursive representations for c_t and k_{t+1} by substituting (4.17) into the first and third equations in (4.12). It can be verified by the reader that these recursive representations are the same as those reported in (3.3) for the special case in which $B(z)$ is 1.

What is of interest to us is the implied moving-average representation for $p_t - p_{t-1}$ in terms of current and past values of w_t. To deduce this representation, note from (4.17) that the one-step-ahead forecast error in mp_t is given by

$$mp_t - mp_{t-1} = -\epsilon(1-\lambda)\,[\rho(\lambda)/\gamma(\lambda)]w_t \,, \tag{4.19}$$

because c_{t-1} and k_t are in J_t. Differencing (4.11), we have that

$$B(L)\,[c_t - c_{t-1}] = [\epsilon(1-\lambda)B(\lambda)^{-1'}\, \mu\rho(\lambda)/\gamma(\lambda)]w_t \,. \tag{4.20}$$

We derive a moving-average representation for $c_t - c_{t-1}$ in a stochastic steady state by applying the inverse operator $B(L)^{-1}$:

$$c_t - c_{t-1} = B(L)^{-1}[\epsilon(1-\lambda)B(\lambda)^{-1'}\, \mu\rho(\lambda)/\gamma(\lambda)]w_t \,. \tag{4.21}$$

Therefore, in a stochastic steady state $\{c_t - c_{t-1}\}$ depends only on the scalar noise $\rho(\lambda)w_t$. While the consumption goods are not proportional, the stochastic process $\{c_t - c_{t-1}\}$ is stochastically singular. Premultiplying both sides of (4.21) by μ' gives:

$$p_t - p_{t-1} = \psi(L)\pi w_t \,, \tag{4.22}$$

where

$$\begin{aligned} \psi(z) = \mu' B(z)^{-1} B(\lambda)^{-1'} \mu / [\mu' B(\lambda)^{-1} B(\lambda)^{-1'} \mu] \\ \text{and} \;\; \pi = (1-\lambda)\rho(\lambda)/\gamma(\lambda) \,. \end{aligned} \tag{4.23}$$

Notice that $\psi(\lambda) = 1$ and that π is the same as in martingale model of Section 3. The scalar lag polynomial $\psi(L)$ occurs in (4.23) because of the nonseparabilities over time in the induced preferences for the payout. In contrast to Section 3, $\{p_t\}$ will not be a martingale unless $\psi(z) = 1$ for all z. Since A3 is maintained, $\rho(\lambda)$ and hence π is different from zero. This eliminates the possibility that the payout process is deterministic. As in Section 3, $\{p_t - p_{t-1}\}$ depends only on current and past values of a scalar noise πw_t.

To relate this model to the analysis in Section 2, we assume that an econometrician observes a process $\{y_t\}$ that satisfies A1 and A2. As in Section 3, the first component of y_t is $s_t - s_{t-1}$ and the second component is $\gamma(L)r_t$. While the econometrician is assumed to observe the payout process $\{p_t\}$, observations on the vector process $\{c_t\}$ of components are not used in the analysis.[15] As in Section 3, we let

$$s_t - s_{t-1} = \sigma(L)w_t \; . \tag{4.24}$$

In light of (4.22) and (4.23), we impose the following restriction on σ:

Restriction R5: $\sigma(z) = (1-z)\rho(z)/\gamma(z) - \psi(z)\pi$ for some scalar $\psi(z)$ satisfying $\psi(z) = \sum_{j=0}^{\infty} \psi_j z^j$, $\sum_{j=0}^{\infty} |\psi|^2 < \infty$ and $\psi(\lambda) = 1$ and for some row vector π.

Notice that R5 is weaker than R3 used in Section 3. Restrictions R1 $[\sigma(\lambda) = 0]$ and R5 together imply that

$$\pi = (1-\lambda)\rho(\lambda)/\gamma(\lambda) \; . \tag{4.25}$$

We now investigate whether (4.25) is a testable restriction. To address this question, it is convenient to use the transformations given in (3.10) and to partition σ^+ and ρ^+ in the same manner. This results in

$$\begin{aligned} s_t - s_{t-1} &= \sigma_1^+(L)w_{1t}^+ + \sigma_2^+(L)w_{2t}^+ \\ \gamma(L)r_t &= \rho_1^+(L)w_{1t}^+ + \rho_2^+(L)w_{2t}^+ \\ p_t - p_{t-1} &= |\pi|\psi(L)w_{1t}^+ \; . \end{aligned} \tag{4.26}$$

Restriction (4.25) implies that

$$(1-\lambda)\rho_1^+(\lambda)/\gamma(\lambda) = |\pi| \tag{4.27}$$

and, by construction,

$$\rho_2^+(\lambda) = 0 \; . \tag{4.28}$$

First, we investigate whether (4.27) is testable. To address this question we decompose y_t as

$$y_t = y_{1t} + y_{2t} \tag{4.29}$$

where

$$y_{1t} = C_1^+(L)w_{1t}^+ \quad \text{and} \quad y_{2t} = C_2^+(L)w_{2t}^+ \ . \tag{4.30}$$

This decomposition can be constructed as follows. Note that linear combinations of current, past and future values of w_{1t}^+ generate the same Hilbert space as linear combinations of current, past and future values of $p_t - p_{t-1}$. Also note that since w_{2t}^+ is uncorrelated with w_{1t}^+, y_{1t} is the least squares regression of y_t onto current, past and future values of w_{1t}^+ and hence onto current, past and future values of $p_t - p_{t-1}$. This means that y_{2t} is the vector of regression errors. The function $\psi(z)$ may have zeros inside the unit circle of the complex plane.[16] As a consequence, linear combinations of current and past values of w_{1t}^+ may generate a strictly larger Hilbert space than linear combinations of current and past values of $p_t - p_{t-1}$. As such, the receipt process may Granger-cause the payout process in this model. For this reason, it is potentially important that future values of $p_t - p_{t-1}$ be included in the regression when forming y_{1t}.

In contrast to Section 3, the scalar noise w_{1t}^+ is not identifiable. Instead, we form a Wold representation of the stochastically singular (rank one) process $\{y_{1t}\}$,

$$y_{1t} = C_1^*(L)w_{1t}^* \ , \tag{4.31}$$

where $\{w_{1t}^*\}$ is scalar white noise with a unit variance, and where linear combinations of current and past values of w_{1t}^* generate the same Hilbert space as linear combinations of current and past values of y_{1t}.

As in the analysis in Section 2, we know that $\{w_{1t}^*\}$ and $\{w_{1t}^+\}$ must be related via

(4.32)

$$w_{1t}^+ = D_1(L^{-1})w_{1t}^* \quad \text{where} \ D_1(z) = \sum_{j=0}^{\infty} d_j(z)^j \ , \ \sum_{j=0}^{\infty} |d_j|^2 < \infty$$

and

$$D_1[\exp(i\theta)]D_1[\exp(-i\theta)] = 1 \quad \text{for almost all } \theta \ . \tag{4.33}$$

Therefore,

$$C_1^+(z) = C_1^*(z)D_1(z)\ . \tag{4.34}$$

The function $D_1(z)$ can have zeros inside the unit circle but it cannot have a zero at λ, because under A3, $|\pi|$ is different from zero and $\psi(\lambda)$ is one. Hence it is not possible for the first two entries of $C_1^+(\lambda)$ both to be zero under R5. The same must be true for $C_1^*(z)$. Let $\sigma_1^*(z)$ denote the first row of $C_1^*(z)$. An implication of (4.27) and (4.34) is

Restriction R6: $\sigma_1^*(\lambda) = 0$.

Since $\sigma_1^*(z)$ is identifiable up to a sign, R5 is testable given knowledge of the discount factor λ.

Proposition 6: Suppose that A1–A4 are maintained. Then R1 and R5 imply R6.

While R6 is designed to accommodate nonseparabilities in preferences, in general, neither $B(z)$ nor $\psi(z)$ can be identified. In contrast to Section 3, the noise w_{1t}^+ is not necessarily identified. Nevertheless the theoretical model imposes enough structure via R5 to guarantee that R1 is testable.

Finally, we investigate whether R6 exhausts all of the testable implications. This amounts to assessing the empirical content of (4.28) $[\rho_2^+(\lambda) = 0]$. Here we simply mimic the logic of Section 3 to conclude:

Proposition 7: Suppose that A1–A5 are maintained. If R5 and R6 are satisfied, then it is always possible to find a moving-average representation $y_t = \hat{C}(L)\hat{w}_t$ that satisfies R1.

The first entry of $\hat{w}_t$ can be chosen to be w_{1t}^* and the remaining components constructed as described in Sections 2 and 3.

Recall from (4.21) that the theoretical model studied in this section implies that the consumption goods all depend on a scalar noise process $\{\pi w_t\}$. In the analysis so far, we have assumed that the econometrician uses data only on total consumption $\{p_t - p_{t-1}\}$. An alternative approach is to assume that the components of consumption are measured with error and to model the observed income and consumption components as a dynamic factor model of the sort studied by Geweke (1977b), Sargent and Sims (1977), Geweke and Singleton (1981a, 1981b), Engle and Watson (1981) and Watson and Engle (1983) with a single factor. The present-value budget balance restriction would then apply to the single factor. The analysis in this section can be modified appropriately to apply to such a model.[17]

5. An Empirical Example

In this section we test R6 using aggregate post war US time series on consumption and labor income. We describe in turn the data, the parameterization for the underlying time series, the estimation method, and the empirical results.

5a. Data

We used aggregate data on total consumption and labor income from 1953:2 - 1984:4 supplied to us by Kenneth West. The original source for the labor income series is Blinder and Deaton (1985). West (1988) used consumption of nondurables and services excluding clothing and services. For this series to be comparable to the labor income series, West scaled the consumption series to reflect the fact that his measure of nondurable consumption is only a portion of total consumption. An advantage of proceeding in this fashion would be that preferences for this nondurable consumption good may be modeled plausibly as time separable. The analysis in Section 3 could then be exploited in studying present-value budget balance. We adopted a somewhat different approach. We used data on total consumption and tested restriction R6 derived in Section 4. Recall that the economic model proposed in Section 4 permits the *indirect* preferences for total consumption to be nonseparable over time. For this reason we were not compelled to remove the consumption of durable goods from total consumption.

5b. Parameterization

We modeled the bivariate consumption and income process using a two-factor specification. These factors are identified by assuming that the factors are mutually uncorrelated and that the second factor has no impact on consumption. In Section 4, we showed that a factor decomposition can always be obtained by representing the rank one process $\{y_{1t}\}$ in (4.29) and (4.30) in terms of a scalar time series, say $\{f_{1t}\}$.

For convenience, we model each factor as having an Nth-order univariate autoregressive representation:

$$f_{jt} = b_j(L)f_{jt-1} + v_{jt} \tag{5.1}$$

where $[1 - zb_j(z)]$ is an $(N+1)^{\text{th}}$-degree polynomial with zeros that are outside the unit circle of the complex plane. Total consumption is then modeled as a distributed lag of $\{f_{1t}\}$ and labor income as distributed lags of both $\{f_{1t}\}$ and $\{f_{2t}\}$:

$$\begin{bmatrix} p_t - p_{t-1} \\ r_t - r_{t-1} \end{bmatrix} = \begin{bmatrix} a_{11}(L) & 0 \\ a_{21}(L) & a_{22}(L) \end{bmatrix} \begin{bmatrix} f_{1t} \\ f_{2t} \end{bmatrix} \tag{5.2}$$

where $a_{ij}(z)$ is a Nth-degree polynomial. The shock process $\{v_t\}$ is serially uncorrelated with $E(v_t v_t') = I$.

To relate this setup to the analysis in Section 4, it is convenient to deduce the implied moving-average average representation for $\{y_t\}$ in terms of $\{v_t\}$. It follows from (5.1) and (5.2) that

(5.3)
$$\begin{aligned} p_t - p_{t-1} &= \Big\{a_{11}(L)/[1 - Lb_1(L)]\Big\}v_{1t} \\ r_t - r_{t-1} &= \Big\{a_{21}(L)/[1 - Lb_1(L)]\Big\}v_{1t} + \Big\{a_{22}(L)/[1 - Lb_2(L)]\Big\}v_{2t}\ . \end{aligned}$$

Subtracting the first equation in (5.3) from the second gives:

(5.4)
$$s_t - s_{t-1} = \Big\{[a_{11}(L) - a_{21}(L)]/[1 - Lb_1(L)]\Big\}v_{1t} + \Big\{a_{22}(L)/[1 - Lb_2(L)]\Big\}v_{2t}.$$

The moving-average representation given by (5.3) and (5.4) coincides with (4.26) when $v_{jt} = w_{jt}^{+}$. In this case

$$\begin{aligned} \sigma_1^{+}(z) &= [a_{11}(z) - a_{21}(z)]/[1 - zb_1(z)] \\ \rho_1^{+}(z) &= a_{21}(z)/[1 - zb_1(z)] \\ \sigma_2^{+}(z) &= \rho_2^{+}(z) = a_{22}(z)/[1 - zb_2(z)] \end{aligned} \tag{5.5}$$

Also $\psi(z)$ is proportional to $a_{11}(z)[1 - z\,b_1(z)]$. Recall that the time nonseparabilities in the induced preference ordering for $\{p_t\}$ are reflected in $\psi(z)$. Present-value budget balance implies that $\sigma_j^{+}(\lambda) = 0$ for $j = 1, 2$ which, in this case, is equivalent to

$$a_{11}(\lambda) = a_{21}(\lambda) \tag{5.6}$$

and

$$a_{22}(\lambda) = 0\ . \tag{5.7}$$

For the reasons given in Section 4, it is not necessarily true that $v_{jt} = w_{jt}^{+}$. Furthermore, the shocks w_{jt}^{+} are not identifiable without further restrictions. Recall from Section 4 that the inability to identify w_{2t}^{+} makes a test based on (5.7) uninteresting. Relation (5.6) is, however, potentially testable because the fundamental shock process $\{w_{1t}^{*}\}$ for $\{y_{1t}\}$ is identifiable. Furthermore, as long as $a_{11}(z)$ and $a_{21}(z)$ do not have common zeros, w_{1t}^{*} is equal to v_{1t} up to a sign convention. In this case (5.6) is equivalent to the Restriction R6. Thus we make (5.6) the focus point of our empirical analysis.

5c. Estimation Method

We tested (5.6) using the method of maximum likelihood assuming a Gaussian likelihood function. In evaluating the likelihood function, the time t components of both series were scaled by μ^{-t} where the growth rate $\log(\mu)$ was estimated. First-differences of the resulting series were taken and sample means removed. The scaled series were modeled as stationary processes using the two factor specification just described. We used a transformation suggested by Monahan (1984) to ensure that $b_j(z)$ has zeros outside the unit circle of the complex plane for $i = 1, 2$. The likelihood for the stationary model was evaluated using recursive state-space methods as described in Chapter 3. An extra (Jacobian) factor was included in the likelihood function to accommodate the scaling of the original time series.

5d. Empirical Results

The unrestricted model was fit for $N = 1, 2, 3$. The log-likelihood values for these cases are reported in Table 1.

Table 1: Log-Likelihood Values for
the Unrestricted Parameterization

N	log-likelihood value
1	−897.21
2	−888.58
3	−878.92

Increasing N by one introduces five new parameters into the model. As seen in Table 1, the changes in the values of log-likelihood function are substantial from the vantage point of the likelihood ratio statistic. On the other hand, from the vantage point of the Schwarz (1978) criterion for model selection, these changes are not substantial. The Schwarz criterion was developed for a different estimation environment and is known to be conservative. Under a nonparameteric perspective in which a finite-dimensional parameterization is viewed as an approximation to an infinite parameter model, it is not apparent that model selection based on either classical likelihood ratio inference or the Schwarz criterion can be justified. We did not explore larger values of N for reasons of numerical tractability and also because the analysis in the next section suggests that restriction (5.6) has very little content for generous parameterizations.

For each N, Table 2 gives the fraction of variation of each series that is explained by the first factor. Recall that, by construction, this fraction is one for consumption.

Table 2: Fraction of Variation Explained by the First Factor for the Unrestricted Parameterization

N	net surplus	labor income
1	.17	.47
2	.22	.53
3	.25	.51

Notice that for each of the parameterizations, about half of the variation in labor income can be attributed to the first factor. It is this variation that forms the basis for our test of present-value budget balance.

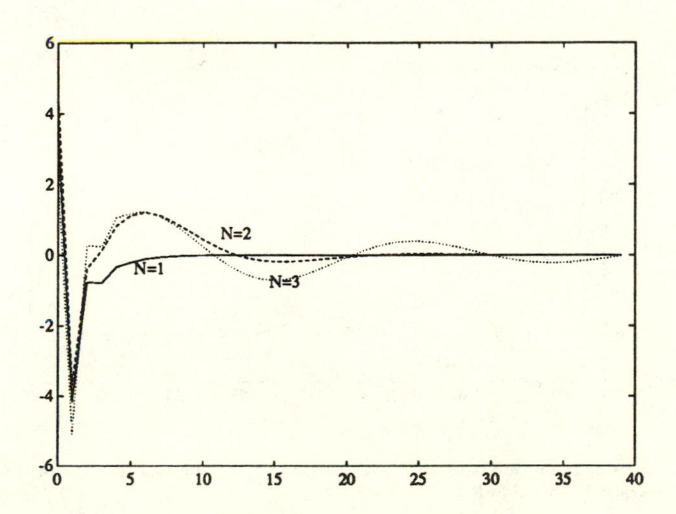

Figure 1. Impulse Responses of Net Surplus in the Unrestricted Models, N = 1, 2, 3.

Figure 1 gives the estimated impulse response function for the first difference of the net surplus process to a surprise movement in v_{1t}

for each of the three specifications of N. In other words, this figure reports the estimated coefficients of the power series expansion of $[a_{11}(z) - a_{21}(z)]/[1 - zb_1(z)]$. The pattern of the response is similar across the three different specifications. In all cases the original increase in the net surplus is more than offset after one time period. Figure 2 gives the estimated impulse response function for the first-difference in consumption and labor income separately for the $N = 3$ run. The peak responses for consumption and labor income both occur after one time period. The nontrival response of consumption to the first shock suggests that it is important to allow for nonseparablities in preferences. In fact, the coefficients of the power series expansion of $\psi(z)$ are proportional to the impulse response function for $\{p_t - p_{t-1}\}$. The similar response patterns of consumption and income might be taken to suggest a trivial model in which consumption and income are the same. Such an interpretation would be misleading, however, because the first shock accounts for all of the variability in consumption but only half of the variability in income.

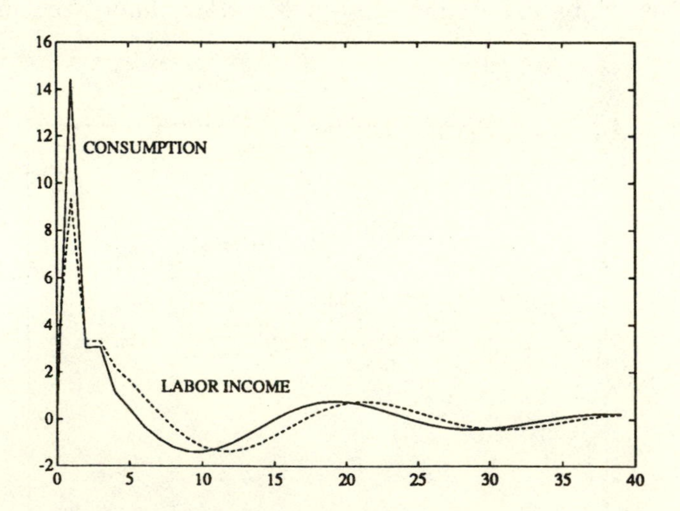

Figure 2. Impulse Response of Consumption and Labor Income in the Unrestricted N = 3 Model.

Restricted versions of the model were also estimated for $N = 1, 2, 3$. For these versions, the $a_{11}(z)$ polynomial was parameterized as follows. First we constructed the function

$$\psi(z) = \phi(z)/[1 - zb_1(z)] \tag{5.8}$$

where $\phi(z)$ is an Nth-degree polynomial. In order that $\psi(\lambda) = 1$, we required that

$$\phi(\lambda) = 1 - \lambda b_1(\lambda) \ . \tag{5.9}$$

The polynomial ϕ was parameterized as

$$\phi(z) = \sum_{j=0}^{N} \phi_j z^j \ , \tag{5.10}$$

where the ϕ_j's were treated as free parameters for $j = 1, 2, \ldots, N$ and ϕ_0 was chosen so as to satisfy (5.9). Finally, the polynomial $a_{11}(z)$ was constructed as

$$a_{11}(z) = a_{21}(\lambda)\phi(z)/[1 - \lambda b_1(\lambda)] \ . \tag{5.11}$$

Note that (5.6) is satisfied by construction, and that there are only N underlying parameters of $a_{11}(z)$, whereas in the unrestricted estimation there were $N + 1$ such parameters.

We estimated the restricted model for five different values of $\log(\lambda)$ ranging from $-.005$ to $-.025$.[18] The values of these likelihoods are reported in Table 3:

Table 3: Log-Likelihood Values for the Restricted Parameterization

			$-\log\lambda$		
N	.005	.010	.015	.020	.025
1	−897.76	−897.73	−897.72	−897.71	−897.69
2	−889.69	−889.69	−889.69	−889.68	−889.68
3	−879.02	−879.02	−879.02	−879.02	−879.02

As is evident from Table 3, the restricted log-likelihood function is not very sensitive to the pre-specified choice of $\log(\lambda)$. Comparing the restricted likelihood values in Table 3 to the unrestricted likelihood values in Table 4, the evidence against the present-value-budget-balance restriction is very weak. In estimating the unconstrained ($N = 3$) model we were unable to find likelihood values that showed any appreciable improvement over the constrained value. Notice that for $N = 3$ the unconstrained likelihood value is only .10 higher than the constrained value. Thus we find that evidence against permanent-income-type models cannot be attributed to violation of present value budget balance.

Table 4 reports the fraction of variance explained by the first factor. Since these numbers are insensitive to the choice of $\log(\lambda)$, only numbers for $\log(\lambda) = -.015$ are reported.

Table 4: Fraction of Variation Explained by the First Factor for the Restricted Parameterization ($N = 3$)

N	net surplus	labor income
1	.16	.52
2	.22	.53
3	.19	.51

Not surprisingly these results are very similar to those reported in Table 2.

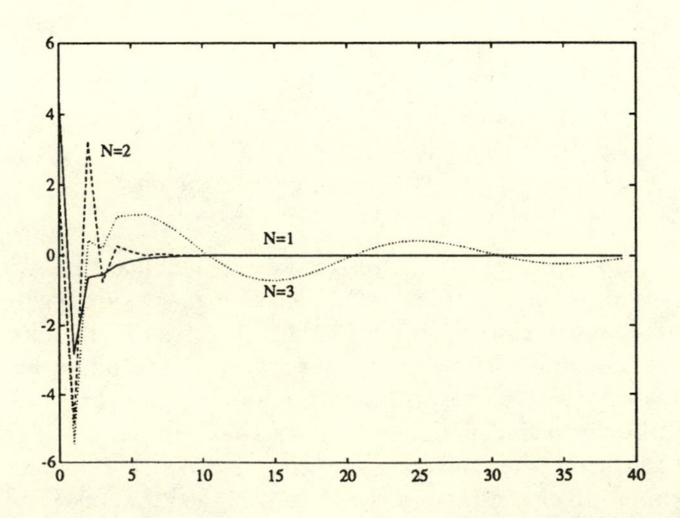

Figure 3. Impulse Responses of Net Surplus in the Restricted Models, N = 1, 2, 3.

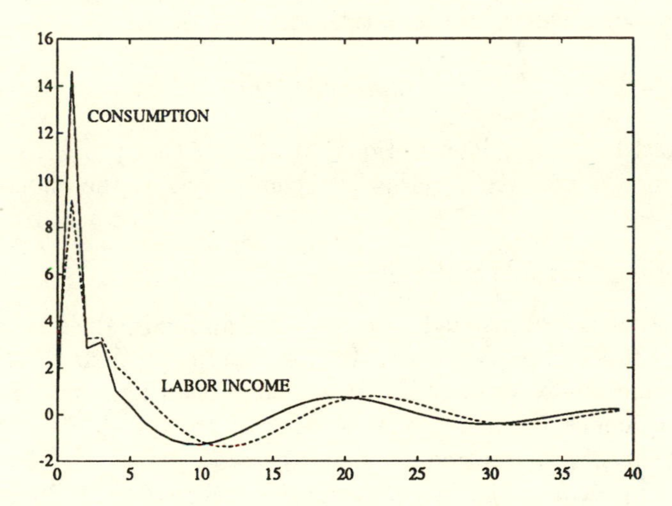

Figure 4. Impulse Responses of Consumption and Labor Income in the Restricted N = 3 Model.

Figure 3 gives the impulse response functions for net surplus and Figure 4 gives the impulse response functions for consumption and labor income for the restricted [$\log(\lambda) = -.015$] version of the model with $N = 3$. These figures are the restricted counterparts to Figures 1 and 2 respectively. Figures 2 and 4 are remarkably similar for the reasons described previously.

6. Approximation Error

In this section we investigate the effect of model approximation error when testing restriction R6. For simplicity, we focus on the case in which $\{y_t\}$ is two-dimensional. Recall that the payout series is a linear combination of the net surplus series and the receipt series. In this section, we find it convenient to make $(1-L)p_t$ rather than $(1-L)s_t$ the first component of y_t. The moving-average representation for $\{y_t\}$ is denoted:

$$y_t = \begin{bmatrix} p_t - p_{t-1} \\ r_t - r_{t-1} \end{bmatrix} = C(L)w_t^+ \tag{6.1}$$

Consistent with A3 and R5 for $\gamma(z) = 1 - z$, we assume that $C(z)$ is lower triangular:

$$C(z) = \begin{bmatrix} \pi_1(z) & 0 \\ \rho_1(z) & \rho_2(z) \end{bmatrix} . \tag{6.2}$$

We impose an additional assumption that accommodates a rich collection of moving-average representations:

Assumption A5: $(1/2\pi)\int_{-\pi}^{\pi}\left\{|\rho_1[\exp(-i\theta)]|^2/|\rho_2[\exp(-i\theta)]|^2\right\}d\theta < \infty.$

In contrast to the previous sections, we assume that present-value budget balance is violated for the first component of the shock process $\{w_t^+\}$:

Assumption A6: $\pi_1(\lambda) \neq \rho_1(\lambda)$.

Our goal is to construct a sequence of approximating models that converge, in some precise sense, to the true model such that each approximating model satisfies R6. If we can construct such an approximating sequence, it will seriously undermine the testability of R6 when (even small) approximation errors are present.

The approximating models we consider all have the following structure. The $C(z)$ matrix function is replaced by an approximating function that is also lower triangular:

$$C^a(z) = \begin{bmatrix} \pi_1(z)b(z) & 0 \\ \rho_1(z)b^*(z) & \rho_2(z) \end{bmatrix} \tag{6.3}$$

where

$$b(z) = (z-\zeta)/(1-z\zeta) \quad \text{and} \quad b^*(z) = (z-\zeta^*)/(1-z\zeta^*) \tag{6.4}$$

for real numbers ζ and ζ^* in the interval $(0,1)$. Consistent with the approximation criterion that is implicit in maximum likelihood estimation, we use an information measure of the magnitude of the approximation error. Let S^a be the spectral density of the approximating model and let S be the spectral density of the true model. Our measure of approximation error uses the population counterpart to the Gaussian log-likelihood function and has a frequency domain representation:

$$\begin{aligned} \eta(S^a, S) = {} & (1/2\pi)\int_{-\pi}^{\pi}\left\{\log\det[S(\theta)] - \log\det[S^a(\theta)]\right\}d\theta \\ & + (1/2\pi)\int_{-\pi}^{\pi} \text{trace}\,[I - S(\theta)S^a(\theta)^{-1}]d\theta \,. \end{aligned} \tag{6.5}$$

Notice that $\eta(S,S)$ is zero. In addition, it can be shown that $\eta(S^a,S)$ is strictly positive unless $S^a(\theta)$ is equal to $S(\theta)$ for almost all θ.

Given an approximating model satisfying (6.3), representation (6.5) simplifies considerably. Recall that

$$S(\theta) = C[\exp(-i\theta)]\; C[\exp(i\theta)]' \tag{6.6}$$
$$= \begin{bmatrix} |\pi_1(z)|^2 & \rho_1(z^{-1})\pi_1(z) \\ \rho_1(z)\pi_1(z^{-1}) & |\rho_1(z)|^2 + |\rho_2(z)|^2 \end{bmatrix}$$

for $z = \exp(-i\theta)$. Similarly,

(6.7)
$$S^a(\theta) = C^a[\exp(-i\theta)]\; C^a[\exp(i\theta)]'$$
$$= \begin{bmatrix} |\pi_1(z)|^2 & \rho_1(z^{-1})\pi_1(z)b^*(z^{-1})b(z) \\ \rho_1(z)\pi_1(z^{-1})b^*(z)b(z^{-1}) & |\rho_1(z)|^2 + |\rho_2(z)|^2 \end{bmatrix}$$

since $|b(z)|^2 = |b^*(z)|^2 = 1$ for all $|z| = 1$. Note that the matrices $S(\theta)$ and $S^a(\theta)$ have the same determinants and that

$$\begin{aligned} &\text{trace}\,[I - S(\theta)S^a(\theta)^{-1}] = \\ &\quad |\rho_1(z)\pi_1(z^{-1})|^2\;[1 - b^*(z)b(z^{-1})]/\det[S(\theta)] \;+ \\ &\quad |\rho_1(z)\pi_1(z^{-1})|^2\;[1 - b^*(z^{-1})b(z)]/\det[S(\theta)] \\ &= |1 - b^*(z)b(z^{-1})|^2\;|\rho_1(z)|^2/|\rho_2(z)|^2 \end{aligned} \tag{6.8}$$

for $z = \exp(-i\theta)$. Substituting into (6.5), it follows that the magnitude of the approximation error is

$$\begin{aligned} &\eta(S^a, S) = \\ &(1/2\pi)\int_{-\pi}^{\pi}\Big(|1 - \big\{b^*[\exp(-i\theta)]\,b[\exp(i\theta)]\big\}|^2 \\ &\qquad |\rho_1[\exp(-i\theta)]|^2/|\rho_2[\exp(-i\theta)]|^2\Big)\,d\theta\,. \end{aligned} \tag{6.9}$$

Formula (6.9) is very similar in structure to one derived in Sims (1972a) for least squares estimation with strictly exogenous regressors. Notice that the magnitude of the approximation error is a weighted integral of the squared error in approximating 1 by the product $b^*\,[\exp(-i\theta)]$ $b[\exp(i\theta)]$. In light of Assumption A5, the density that weights this squared error induces a finite measure on $(-\pi, \pi]$. Among other things, this assumption guarantees that the approximation error is finite for members of our class of approximating models.

We now construct a sequence of approximating models $\{C_j\}$ for which $\{\eta(C_j, C)\}$ converges to zero. The matrix C_j is given by the right side of (6.3) with b_j replacing b and b_j^* replacing b^* where

$$b_j(z) = (z - \zeta_j)/(1 - z\zeta_j) \text{ and } b_j^*(z) = (z - \zeta_j^*)/(1 - z\zeta_j^*) \tag{6.10}$$

for sequences of real numbers $\{\zeta_j\}$ and $\{\zeta_j^*\}$ in the interval $(0, 1)$. For each j the pair (ζ_j, ζ_j^*) is chosen so that the present-value budget balance restriction R6 is satisfied. We focus on the special case in which $\pi_1(\lambda)$ and $\rho_1(\lambda)$ are both different from zero. Arguments for the other cases proceed in a similar but slightly more complicated fashion.[19]

Let $\{\zeta_j\}$ be a sequence of real numbers in the interval $(0, 1)$ that are distinct from λ but that converge to λ. Note that $\pi_1(\lambda)\, b_j(\lambda)$ is different from zero but that the sequence $\{\pi_1(\lambda)\, b_j(\lambda)\}$ converges to zero. Next choose ζ_j^* so that

$$\pi_1(\lambda)\, b_j(\lambda) = \rho_1(\lambda)\, b_j^*(\lambda) \tag{6.11}$$

It follows from the analysis in Section 4 that (6.11) implies that restriction R6 is satisfied for each of the approximating models. Viewing (6.11) as an equation in ζ_j^*, it follows that ζ_j^* must satisfy

$$\pi_1(\lambda)\, b_j(\lambda)\, (1 - \lambda\zeta_j^*) = \rho_1(\lambda)\, (\lambda - \zeta_j^*)\ . \tag{6.12}$$

For sufficiently large j, equation (6.12) determines ζ_j^* uniquely because $\{b_j(\lambda)\}$ converges to zero and $\rho_1(\lambda)$ is different from zero. Solving for ζ_j^* gives

$$\zeta_j^* = [\pi_1(\lambda)\, b_j(\lambda) - \rho_1(\lambda)\lambda]/[\lambda\pi_1(\lambda)\, b_j(\lambda) - \rho_1(\lambda)] \tag{6.13}$$

Since $\{b_j(\lambda)\}$ converges to zero, $\{\zeta_j^*\}$ converges to λ. By omitting some of the initial entries of the sequence of $\{\zeta_j\}$ and hence also of $\{\zeta_j^*\}$, we can guarantee that all entries of both sequences are in the interval $(0, 1)$.

It remains to show that the approximation error $\eta(C_j, C)$ can be made arbitrarily small. Since $\{\zeta_j\}$ and $\{\zeta_j^*\}$ both converge to λ, $\{b_j^*(z)\, b_j(z^{-1})\}$ converges to one for all $|z| = 1$. Furthermore,

$$\begin{aligned} &|b_j^*(z)|\ |b_j(z^{-1})| \leq \\ &[(1 + \zeta_j)\,(1 + \zeta_j^*)]/[(1 - \zeta_j)\,(1 - \zeta_j^*)] \quad \text{for} \quad |z| = 1\ . \end{aligned} \tag{6.14}$$

The inequality permits us to apply the Dominated Convergence Theorem and conclude that $\{\eta(S, S_j)\}$ converges to zero. We have established:

Proposition 8: Suppose Assumptions A1–A6 are satisfied. Then there exists a sequence $\{S_j\}$ of spectral density functions satisfying R6 such that $\{\eta(S_j, S)\}$ converges to zero.

In light of Proposition 8, the empirical content of Restriction R6 is tenuous. Even when the restriction is not satisfied, it is possible to construct a sequence of models, each of which satisfies R6, that approximate the original model arbitrarily well. These approximating models have the feature that the corresponding sequences $\{\rho_1^j\}$ and $\{\pi_1^j\}$ have zeros that are arbitrarily close to λ. To avoid the negative implications of Proposition 8, one must be able to rule out zeros of these functions in a neighborhood of λ.[20] Recall that in Section 5 we assumed that $\rho_1(z)$ and $\pi_1(z)$ are ratios of finite-order polynomials. If the orders of these polynomials are chosen to be too large, then a finite-parameter version of Proposition 7 holds. Of course it is very difficult in practice to determine the precise order of numerator and denominator polynomials. Furthermore, it is often the case that the parameterizations used in estimation are best thought of as approximations to models with more complicated time series correlation structure. Hence, it seems hard to circumvent the negative conclusion of Proposition 8.

7. Variable Real Interest Rates

In this section, we permit uncertainty in the real interest rate δ of (2.1). We rewrite (2.1) as

$$k_{t+1} = (1 + \delta_t)\, k_t + s_t \quad \text{for} \quad t = 0, 1, 2, \ldots \,. \tag{7.1}$$

Let J_{t-1} denote information available to agents at time $t-1$. Throughout this section, we require

$$E\,(\delta_t \mid J_{t-1}) = \delta \,, \tag{7.2}$$

so that the conditional expectation of interest rates is constant over time. Since $\{\delta_t\}$ is no longer deterministic, the martingale implications described in Sections 2 and 3 may no longer be applicable.

To deduce testable restrictions, we take expectations of (7.1) conditioned on time $t-1$ information. This yields

$$E(k_{t+1} \mid J_{t-1}) = (1 + \delta)\, k_t + E(s_t \mid J_{t-1})\,, \; t = 0, 1, \ldots \,, \tag{7.3}$$

with k_0 given, since k_t is determined at time $t-1$. We impose assumption A1 and terminal condition

$$E \sum_{t=0}^{\infty} \lambda^t\, (k_t)^2 < +\infty \,,$$

where $\lambda = 1/(1+\delta)$. Solving (7.3) forward gives

$$(7.4) \qquad k_t = -\sum_{j=0}^{\infty} \lambda^{j+1} E\left(s_{t+j} \mid J_{t-1}\right) \quad , \quad t = 0, 1, \ldots .$$

For reasons similar to those given in Section 2, (7.4) imposes no testable restrictions on the joint process $\{(r_t, p_t)\}$ beyond those described in Section 2. However, (7.4) does have interesting implications for the question studied by Hamilton and Flavin (1986), namely the restrictions imposed by present value budget balance on the joint process $\{(k_{t+1}, r_t, p_t)\}$. In fact (7.4) is a version of the class of present value models investigated by Campbell and Shiller (1987).

We continue to assume that information available to agents has the structure described in Section 2. In particular, recall that s_t is the first component of an m-dimensional vector process y_t that has representation

$$y_t = C(L)\, w_t \ ,$$

where w_t is the vector of innovations to agents' information. The first row of the above system of equations is

$$s_t = \sigma(L)\, w_t \ .$$

Using the above formula together with results of Hansen and Sargent (1980a), permits (7.4) to be represented as

$$(7.5) \qquad k_{t+1} = \kappa(L) w_t \ \text{ where } \ \kappa(z) = -\left[\frac{\lambda\sigma(z) - \lambda\sigma(\lambda)}{(z-\lambda)}\right] .$$

Equation (7.5) translates directly into a restriction on the moving-average representation for $\{(k_{t+1}, y_t')\}$ in terms of $\{w_t\}$.

So long as the dimension n of w_t is greater than the dimension m of y_t, a composite process $\{(k_{t+1}, y_t')\}$ satisfying (7.5) can be stochastically nonsingular. Stochastic nonsingularity can emerge when k_{t+1} cannot be expressed as a linear function of current, past and future values of y_t. In this circumstance, k_{t+1} can reveal additional information about w_t. Indeed, $\{(k_{t+1}, y_t'), (k_t, y_{t-1}') \ldots\}$ could well generate a smaller information set than does $\{w_t, w_{t-1}, \ldots\}$.[21]

Hamilton and Flavin (1986) formulated and tested a version of (7.5) that emerges under the special assumption that s_t is a univariate process with an innovation that reveals w_t, so that s_t is Granger caused by no other variables in $\{(k_{t+1}, y_t')\}$. Under this special assumption, restriction (7.5) implies that the $\{(k_{t+1}, y_t')\}$ process is stochastically singular and that the regression of k_{t+1} on current and lagged s_t fit by Hamilton and Flavin will have an R^2 of unity. If one were to drop the assumption that s_t is Granger caused by no other components of $\{(k_{t+1}, y_t')\}$, a serially correlated *Shiller error* would emerge in the relation fit by Hamilton and Flavin. The presence of that error would require resorting to estimators of the class described by Hansen and Sargent (1982) and Hayashi and Sims (1983) in order to obtain consistent parameter estimates and valid test statistics.

Further insights about efficient estimation and about the structure of the restrictions are gained by noting that the model for $\{(k_{t+1}, y_t')\}$ induced by (7.5) is a member of a class of *exact linear rational expectations models* studied by Hansen and Sargent in Chapter 3. The results in Chapter 3 imply that the restrictions given by (7.5) always apply to a Wold moving-average representation for $\{(k_{t+1}, y_t')\}$.[22] Furthermore, even if the moving-average representation in terms of w_t turns out to be a Wold representation, it is always possible to find some other moving-average representation with the same number of noises that satisfies the restrictions and is not a Wold representation. Thus, unlike the restrictions studied in Section 2, the restrictions given in (7.5) apply to both the Wold moving-average representation and to some other moving-average representations. Evidently, the restrictions (7.5) have a different structure from those studied in Section 2.

Finally, it is straightforward to extend the analysis in this section to accommodate unit roots in the payoff and receipt processes along the lines described in Section 2. The asset process $\{k_t : t = 0, 1, \ldots\}$ will inherit the nonstationarity; however, (7.5) implies that the process $\{k_{t+1}, r_t, p_t) : t = 0, 1, \ldots\}$ will be co-integrated in the sense of Engle and Granger (1987) (e.g. see Chapter 3 and Campbell and Shiller 1987).

8. Conclusions

In Section 2, we supplemented the hypothesis of present-value budget balance with two more hypotheses, the existence of a time invariant moving-average representation for the process $\{(r_t, p_t)\}$ and constancy of the net rate of interest. We found that present-value budget balance is so weak an hypothesis that it imposes literally no observable restrictions on the process $\{(r_t, p_t)\}$. This weakness illustrates how difficult it is to verify a present-value budget constraint that restricts the entire (infinite-dimensional) future of the $\{(r_t, p_t)\}$ process. The restriction acquires content only if the hypothesis of a time-invariant linear representation is supplemented with additional hypotheses that reduce the parameter space in ways that render inadmissible the transformations that we exhibited in Section 2. We showed that the present-value budget balance restriction is delicate in the sense that it cannot hold for a Wold moving-average representation, which is the representation that is typically recovered using vector autoregression methods.

In Sections 3 and 4, we showed that the present value budget balance restriction acquires additional content when it is part of a class of linear-quadratic models of optimal consumption (or optimal tax collections). In those models, present value budget balance imposes an additional restriction on the $\{(r_t, p_t)\}$ process above and beyond those imposed by martingale characterizations stemming from Euler equations. A sense in which even these restrictions are tenuous was described in Section 6.

Section 7 described the restrictions on the joint $\{(k_{t+1}, p_t, r_t)\}$ process imposed by the assumption that $E_{t-1}\delta_t = \delta$ for all t. It turns out that those restrictions deliver an *exact linear rational expectations model*. In the next paper in this volume (Chapter 6) Roberds estimates such a model for post war U.S. time series on government debt and deficits.

Appendix A

In this appendix we show that $D(z)$ given in (2.20) can be used to build a Wold representation for the composite process $\{x_t\}$. Use the Wold representation for $\{y_t\}$, namely,

$$y_t = C^*(L)w_t^* \tag{A.1}$$

to form a two-sided moving-average representation for $\{x_t\}$ by adding a row

$$k_t = \kappa^*(L)w_t^* \ , \qquad \text{where} \quad \kappa^*(z) \equiv -\lambda z\sigma^*(z)/(z-\lambda) \ , \tag{A.2}$$

and $\sigma^*(z)$ is the first row of $C^*(z)$. Use $D(z)$ given in (2.20) to form an alternative moving-average representation for $\{x_t\}$:

$$y_t = C^+(L)w_t^+ \ , \quad k_t = \kappa^+(L)w_t^+ \tag{A.3}$$

where

$$C^+(z) \equiv C^*(z)D(z) \quad \text{and} \quad \kappa^+(z) \equiv \kappa^*(z)D(z) \ . \tag{A.4}$$

Note that

$$\begin{aligned} \kappa^+(z) &= -\lambda z\sigma^*(z)D(z)/(z-\lambda) \\ &= -\lambda z\sigma^*(z)\,Q\begin{bmatrix} 1/(1-\lambda z) & 0 \\ 0 & [1/(z-\lambda)]I \end{bmatrix} . \end{aligned} \tag{A.5}$$

Since $\sigma^*(\lambda)Q$ has zeroes in all entries except the first, $\kappa^+(z)$ has a removable singularity at $z = \lambda$, implying that $\kappa^+(L)$ is one-sided. Therefore, $(A.3)$ gives a one-sided moving-average representation for $\{x_t\}$.

It remains to show that representation $(A.3)$ is a Wold representation. Suppose to the contrary that the rank of the matrix $\begin{bmatrix} C^+(\lambda) \\ \kappa^+(\lambda) \end{bmatrix}$ is m-1. In this case there exists an orthogonal matrix Q^+ such that the first column of $\begin{bmatrix} C^+(\lambda) \\ \kappa^+(\lambda) \end{bmatrix} Q^+$ contains all zeroes. Hence the matrix function

$$(A.6) \qquad \begin{bmatrix} \tilde{C}(z) \\ \tilde{\kappa}(z) \end{bmatrix} \equiv \begin{bmatrix} C^+(\lambda) \\ \kappa^+(\lambda) \end{bmatrix} Q^+ \begin{bmatrix} (1-\lambda z)/(z-\lambda) & 0 \\ 0 & I \end{bmatrix}$$

has a removable singularity at $z = \lambda$, and the composite process $\{x_t\}$ has a one-sided moving-average representation:

$$(A.7) \qquad y_t = \tilde{C}(L)\tilde{w}_t \ , \quad k_t = \tilde{\kappa}(L)\tilde{w}_t$$

for some vector white noise $\{\tilde{w}_t\}$. Consequently, the first row of $\tilde{C}(\lambda)$ satisfies R1. However,

$$(A.8)$$

$$\tilde{C}(z) = C^*(z)Q \begin{bmatrix} (z-\lambda)/(1-\lambda z) & 0 \\ 0 & I \end{bmatrix} Q^+ \begin{bmatrix} (1-\lambda z)/(z-\lambda) & 0 \\ 0 & I \end{bmatrix}$$

implying that $\det[C^*(z)] = \det[\tilde{C}(z)]$. Thus $\tilde{C}(z)$ also satisfies R2 which contradicts Proposition 2. This proves that $(A.3)$ is a Wold representation for $\{x_t\}$.

Notes

1. The recent paper of Trehan and Walsh (1988) studied these questions as well as tax smoothing.

2. More formally, the notation J_t will be used both to denote the sigma algebra generated by w_t, w_{t-1}, ... and the space of all random variables with finite second moments that are measurable with respect to this sigma algebra.

3. For definitions of concepts from the theory of complex variables such as Laurent series, Taylor series, poles and removable singularities see Churchill, Brown and Verhey (1974). For applications of these concepts in macroeconomics, see Hansen and Sargent (1980a), Whiteman (1983) and Sargent (1987b).

4. Sargent (1987a) did not show the sense in which the restriction is vacuous and did not fully explore the implication that the restriction does not apply to a Wold representation.

5. Recall that the white noise sequence in a Wold representation is not necessarily a martingale difference sequence but instead satisfies the weaker requirement of being serially uncorrelated. We do not investigate the implications of the stronger restriction that the vector white noise be a martingale difference sequence.

6. While A2 is essential to the proof of Proposition 2, it is adopted only as a matter of convenience for Proposition 3. We leave it to the reader to show that Proposition 3 holds when only A1 is maintained. Also, several examples in Sargent (1987b, Ch. XIII) purport to show instances in which restriction R1 is testable. These examples all hinge on maintaining a low-dimensional parameterization of C. The reader can verify that these restrictions vanish when the dimensionality of the parameterizations is expanded.

7. The language *receipt* and *payouts* is inspired by the permanent income model of consumption, but is perverse when applied to Barro's tax smoothing model. (See (3.1)–(3.2) in the text.) In Barro's model, what we call *receipts* corresponds to government expenditures, while what we call *payouts* corresponds to total tax collections.

8. In the spirit of this paper, issues that occur when $\beta(z)$ is approximated by ratios of polynomials potentially are relevant to our analysis. Sims (1972a) demonstrated that objects such as sums of lag coefficients are particularly sensitive to errors in approximating $\beta(z)$. The same observation, however, does not apply to approximating $\beta(z)$ for $|z| < 1$.

9. As pointed out to us by Chi Wa Yuen, if λ is not known, then R4 can be used to help identify λ. In general, R4 can only be used to identify λ locally and not globally because the function $(1 - z)\beta(z) - \gamma(z)$ can have multiple zeros in (0,1). A finding of no zeroes in this region constitutes evidence against R4. However, the testable implications of R4 are considerably weaker and the resulting statistical tests harder to implement when λ is not known *a priori*.

10. In Barro's (1979) model, the martingale implication might be construed as applying to tax rates instead of taxes. Let τ_t denote the tax rate. In this case the present value budget balance restriction could be tested by regressing the surplus process onto the current and past values of $(\tau_t - \tau_{t-1})/\tau_{t-1}$ and checking the discounted sum of coefficients.

11. Restriction R4 is absent in the permanent income model studied by Campbell (1987) because k_{t+1} in Campbell's model is not restricted to be in J_t. Instead k_{t+1} also depends on a shock labeled *unanticipated capital gains* that is only observed by economic agents as of time $t + 1$.

12. West (1988) correctly pointed out that obtaining a more powerful test may not be essential in the case of the permanent income model of consumption because evidence using his test is sufficient to challenge the validity of the model.

13. Hansen (1987) showed that specifications in which there are more services than goods can often be converted into specifications in which the number of goods and services are the same without changing the optimal decision rules for consumption. Hence the restriction in A4 that $B(z)$ be a square matrix can often be relaxed. The further restriction that $B(z)$ be nonsingular rules out cases in which *stable* service processes may require *unstable* consumption processes as in, for example, the rational addiction model of Becker and Murphy (1988).

14. Heaton (1989) deduced a continuous time counterpart in a model with a single consumption good but a general specification of the temporal nonseparabilities in preferences.

15. Wilcox (1989) noted that one of the difficulties in NIPA measures of consumption is splitting retail sales into durable and nondurable components. Our model avoids having to make this split.

16. To illustrate that $\psi(z)$ can be zero inside the unit circle of the complex plane, consider the following setup. Suppose that

$$B(z) = \begin{bmatrix} 1 & bz \\ 0 & 1 \end{bmatrix}$$

for $b > 0$ and $\mu' = [1, 1]$. Then

$$\psi(z) = [1 + (1 - bz)(1 - b\lambda)]/[1 + (1 - b\lambda)^2] .$$

It is easy to verify that b can be chosen so that $\psi(z)$ is zero for some $|z| < 1$. For instance, note that

$$\psi(-\lambda) = [2 - (b\lambda)^2]/[1 + (1 - b\lambda)^2] .$$

Then $\psi(-\lambda) = 0$ for $b = \sqrt{2}/\lambda$. An objection to this specification is that in a deterministic steady state, the marginal utility for the first service is negative. This defect can be overcome by premultiplying $B(z)$ by an appropriately chosen orthogonal matrix. This transformation of $B(z)$ does not alter the indirect preferences for consumption goods. In addition, there are initial conditions for the capital stock and a constant endowment sequence for which both consumption goods are strictly positive in a steady state. Such a specification should tolerate at least small amounts of uncertainty in the endowment process and still have the vector of services, consumption goods, and marginal utilities be positive with high probability.

17. An alternative approach suggested by Andy Atkeson is to assume that the indirect preferences for consumption are separable in the first good. This is equivalent to restricting $B(z)$ to have all zeros in the first row and column except for the (1,1) entry. If the first consumption good is measured without error, then no other processes adapted to the information sets of economic agents should Granger-cause the process for the first consumption good. Furthermore, the univariate innovation in this process should reveal

a component of the information set of economic agents to which the present value budget balance restriction applies. To test this restriction still requires that the surplus process also be measured without error. However, it is important to remember that if other consumption goods are also measured without error, the model described in Section 4 implies a stochastic singularity for the observed time series tht is likely to be counter-factual.

18. Recall that the time series are scaled by an estimated growth factor as part of the estimation. The discount factor λ applies to the scaled time series. The estimates of the growth rate $\log(\mu)$ were typically around .0075.

19. For instance, if $\pi_1(\lambda)$ has single zero at λ, then the function $b_j(z)$ should have an additional factor of the form $[(z-\lambda_j)(1-\lambda z)]/[(1-\lambda_j z)(z-\lambda)]$ where $\{\lambda_j\}$ converges to λ. Among other things, this extra factor is designed so that $\pi_1(z)b_j(z)$ has a removable singularity at $z=\lambda$.

20. For the (low order) parameterizations adopted in Section 5, approximation error does not seem to be the source of our nonrejection because $\pi_1(\lambda)$ and $\rho_1(\lambda)$ both seem not to be zero. For both the unrestricted and the restricted estimates, we computed the zeros of $a_{11}(z)$ and $a_{21}(z)$. The only zeros that were similar in magnitude for $a_{11}(z)$ and $a_{22}(z)$ were complex and had absolute values exceeding unity. There were no common zeros in the vicinity of plausible values of λ.

21. This point is related to those that we discuss under the topic of the first of the two difficulties in our paper "Two Difficulties in Interpreting Vector Autoregressions" (Chapter 4). That paper treats a class of examples related to (7.5) in which it can be taken that $n=m$, and in which current and lagged values of the process (k_{t+1}, y_t) spans a smaller information set than does $\{w_t, w_{t-1}, \ldots\}$.

22. Among other things, Campbell and Shiller (1987) exploit this observation and derive the restrictions implied on a finite-order vector autoregression.

6

Implications of Expected Present Value Budget Balance: Application to Postwar U.S. Data

by William ROBERDS

1. Introduction

For time series on the U.S. government budget after World War II, this paper implements the test described in Section 7 of Hansen, Sargent, and Roberds [henceforth HSR], which is Chapter 5 of this book. Recall that Section 7 modifies the setup of earlier sections in two ways. First, interest rates are allowed to be time invariant *ex ante* but not *ex post*. In the notation of HSR, this requires that

$$E\left(\delta_t \mid J_{t-1}\right) = \delta \tag{1.1}$$

where δ_t is the real one period interest rate and J_{t-1} represents information available as of time $t-1$. Second, measures of the debt stock are assumed to be in the econometrician's data set. In Section 2, I summarize how these two assumptions lead to the model formulated in Section 7 of HSR. The model is then tested for postwar U.S. time series on federal government debt and deficits net of interest.

The analysis below is closely related to the work in a number of papers examining the question of *net present value budget balance* using postwar U.S. fiscal data, most notably Hamilton and Flavin (1986).[1] It is also closely related to a number of papers that test *expectational* models of the relationship between stock prices and dividends, as well as that between long rates and short rates, e.g., Campbell and Shiller (1987) and Hansen and Sargent (1981e). Differences and similarities between these papers and the present analysis are noted below.

2. Implications of Expected Net Present Value Budget Balance

As in HSR, let $\{s_t\}$ be a stochastic process of net surpluses, i.e., receipts minus expenditures net of interest. Let $\{k_t\}$ be the stochastic process representing debt at the beginning of period t, denominated in negative dollars when the government is borrowing money. Debt evolves according to the government budget constraint

$$E(k_{t+1} \mid J_{t-1}) = (1+\delta)\,k_t + E(s_t \mid J_{t-1}) \text{ for } t = 0, 1, \ldots \tag{2.1}$$

Replicating the analysis leading up to equation (2.8) of HSR yields the *solution* for k_t

$$k_t = -\sum_{\tau=0}^{\infty} \lambda^{\tau+1} E(s_{t+\tau} \mid J_{t-1}) \tag{2.2}$$

which states that debt must be balanced by the discounted sum of expected future surpluses. Evidently restrictions imposed by (2.2) will be weaker than those implied by equation (2.8) of HSR, which does not contain an expectations operator.[2] Hence the impossibility result (Proposition 2) of HSR does not apply in the present case. To derive the restrictions implied by (2.2), suppose that as in HSR, s_t is contained in an observable vector y_t, and that HSR assumptions A1 (stationarity) and A2 (nonsingularity) hold. Let the equation corresponding to the appropriate row of the MAR for y_t be given by

$$s_t = \sigma(L)\,w_t \tag{2.3}$$

where σ is a one-sided lag polynomial and $\{w_t\}$ is a martingale difference sequence. Applying a prediction formula of Hansen and Sargent (1980) to (2.2) yields a unique one sided representation for $\{k_{t+1}\}$

$$k_{t+1} = \kappa(L)\,w_t \quad \text{where } \kappa(z) = \lambda[\sigma(\lambda) - \sigma(z)]/(z-\lambda)\,. \tag{2.4}$$

Because $\kappa(z)$ is one-sided, equation (2.4) translates directly into restrictions on the moving average representation (henceforth, MAR) of $\{(k_{t+1}, y_t')\}$ in terms of $\{w_t\}$. Recall that w_t represents the innovation to agents' information. So long as the dimension of w_t is greater than the dimension of y_t plus one (i.e., $n+1$ in the notation of HSR), the composite process $\{(k_{t+1}, y_t')\}$ can be nonsingular. Stochastic nonsingularity occurs when the history of the process $\{y_t\}$ generates a strictly smaller information set than does the history of $\{w_t\}$. In this instance,

k_{t+1} can reveal additional information about w_t, implying stochastic nonsingularity of $\{(k_{t+1}, y_t')\}$. The analysis below assumes that such nonsingularity will in fact hold.

The restrictions induced by equation (2.4) are easily tested using a result obtained by Hansen and Sargent (1981e). This result, which applies to *exact* linear rational expectations models such as (2.4), implies that the restrictions given by (2.4) will always apply to the Wold moving average representation for $\{(k_{t+1}, y_t')\}$. Consequently one can replace the polynomials κ and σ in (2.4) with their estimable counterparts κ^+ and σ^+. The restrictions on κ^+ and σ^+ may then be tested using standard methods.

3. Application to U.S. Postwar Data

As an example of how model (2.2) can be applied to data, I estimated a simple version of this model for U.S. quarterly time series over the period 1948Q1–1986Q4. In this application, the dimension of y_t is taken to be one, so that $y_t = s_t$. The real debt series $(-k_t)$ and the real deficit net of interest series $(-s_t)$ are constructed from NIA series. Both series account for the profits of the Federal Reserve System as revenues. Details on the construction of the data series can be found in Appendix A of Miller and Roberds (1987). The series are graphed in Figures 1 and 2.

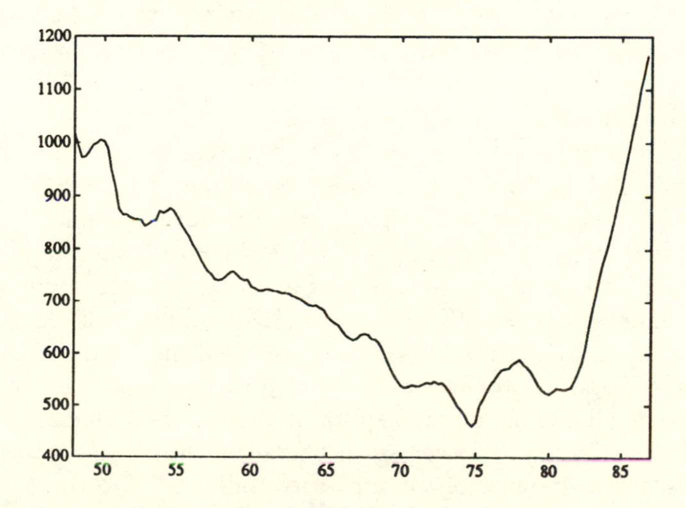

Figure 1. Real Value of Interest Bearing U.S. Federal Debt 1948:1 to 1986:4 (billions of 1982 dollars)

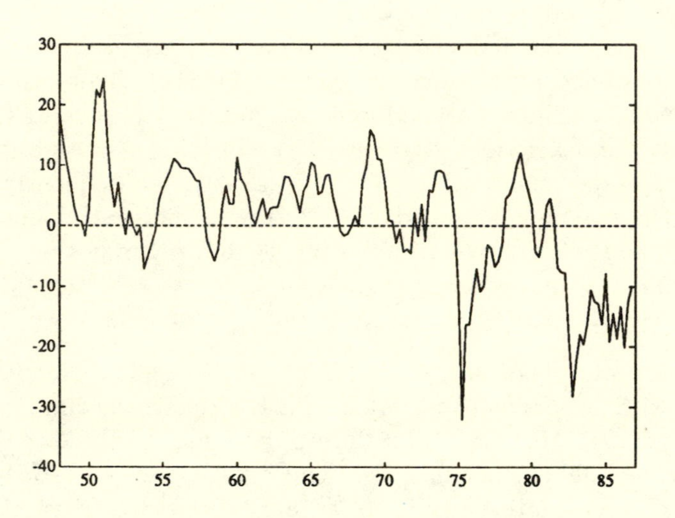

Figure 2. Net of Interest Surplus in U.S. Federal government budget,1948:1 to 1986:4 (billions of 1982 dollars)

Given that the restrictions imposed by (2.4) require stationarity of s_t and hence k_t, some pretesting for nonstationarity is appropriate. Results of standard Dickey-Fuller (henceforth DF) regressions are given in Table 1.

Table 1 shows that the Dickey-Fuller test rejects the null of a unit root for the debt series but not for deficit. These results constitute some prima facie evidence against the validity of the model of Section 2 for postwar U.S. data. If real debt evolves according to equation (2.4), and real deficits (surpluses) are stationary, then real debt cannot be nonstationary. Clearly, the budget cannot be balanced in any meaningful "expected present value" sense if debt diverges over time while deficits continue to fluctuate in a stationary fashion. On the other hand, it may be the case that such pessimistic inferences are unwarranted, because of biases inherent in the DF test. Sims (1988), Sims and Uhlig (1990), DeJong et al. (1988) and others have questioned the applicability of DF and similar classical procedures for determining the presence of a unit root. Specifically, many of the results in these papers suggest that the DF test suffers from low power against near-nonstationary alternatives, leading to a bias in favor of the unit root null.

As an alternative to standard DF tests of a unit root, the papers mentioned above employ Bayesian methods to obtain inferences concerning potential nonstationarity. In Table 2, some of these methods are used to analyze the real debt and deficit series. Following the approach of DeJong and Whiteman (1989 a,b) a sixth order AR model with a constant term was fit to each series. Assuming a normal likelihood function, and diffuse (normal-gamma) prior for the model parameters, Monte Carlo integration was used to obtain estimates of the posterior mean and standard deviation of the modulus of the largest root of the AR polynomial. These estimates are given in Table 2, along with the approximate posterior probability that each of the largest roots is inside the unit circle.

For both series, the Bayesian procedure places most of the posterior probability on stationarity. However, the posterior probability of nonstationarity is much higher (about 16%) for debt than for deficits (less than .1%). Sims' (1988) test of a unit root as a point null slightly favors the unit root over stationarity for debt, and vice versa for deficits. Both inferences, however, could be reversed by a relatively small change in prior odds. On balance, the evidence presented in Table 2 suggests that stationarity is the most likely inference for both series, while difference-stationarity is plausible in the case of debt but somewhat less plausible for deficits. Due to this ambiguity concerning the possible presence of unit roots, two versions of the model were fit to the data. The first was the stationary model described in Section 2. The second model assumes difference-stationarity of the deficit process.

To derive the differenced version of the model, rearrange the terms in equation (2.2) to obtain

$$\delta k_t + s_{t-1} = -\sum_{\tau=0}^{\infty} \lambda^\tau \, E(\Delta s_{t+\tau} \mid J_{t-1}) \tag{3.1}$$

which states that this period's expected deficit including interest payments must be balanced by the discounted sum of expected changes in all future deficits. If $\{\Delta s_t\}$ is taken to be stationary, then equation (3.1) implies that deficits including interest payments, i.e., $\{\Delta k_t\}$ will be stationary. Now assume that HSR assumptions A1 (stationarity) and A2 (nonsingularity) apply to first differences of s_t. Let the MAR for Δs_t be given by

$$s_t = \sigma(L) w_t \tag{3.2}$$

where σ is a one-sided lag polynomial and $\{w_t\}$ is a martingale difference sequence. Applying the Hansen-Sargent (1980a) prediction formula to (3.1) yields a unique one sided representation for $\{\delta k_{t+1} + s_t\}$

$$k_{t+1} = \kappa(L) w_t \quad \text{here } \kappa(z) = [\sigma(\lambda) - \sigma(z)]/(z - \lambda)\,. \tag{3.3}$$

As in the stationary case, equation (3.3) can be directly translated into restrictions on the MAR of $\{[(\delta k_{t+1} + s_t), \ \Delta s_t]\}$. The term $\delta k_{t+1} + s_t$ represents the current period's expectation of the deficit net of interest at the end of next period, i.e., $E(\Delta k_{t+1} \mid J_{t-1})$. Alternatively, this term is proportional to $k_{t+1} + \delta^{-1} s_t$, which represents the discrepancy between the value of debt and its expected net present value, assuming a constant *ex ante* real rate and that s_t follows a random walk. Similar terms appear in the price-dividend/term structure models analyzed by Hansen and Sargent (1981e) and Campbell and Shiller (1987), which are formally identical to the difference stationary model analyzed above.

To implement the tests described above, a bivariate vector autoregression with a constant term was fit to $\{k_{t+1}, \ s_t\}$ for the stationary model and $\{[(\delta k_{t+1} + s_t), \ \Delta s_t]\}$ for the first differences model. Using methods described in Campbell and Shiller (1987), the restrictions implied by (2.4) and (3.3) were reduced to linear restrictions, and tested by means of likelihood ratio tests.[3] Table 3 displays results for both models, under various assumptions about lag lengths and real interest rates.

The results in Table 3 show that the constant *ex ante* real rate model can be rejected at essentially arbitrary significance levels by the postwar U.S. data. Differencing seems to impact little on the significance level of the test statistics. This strong rejection of the model stands in sharp contrast to the findings of Hamilton and Flavin (1986), who conclude that equation (2.2) represents a useful approximation for the postwar U.S. case. Some possible explanations for this discrepancy are considered below.

Hamilton and Flavin's (henceforth HF) study differs from the present one in the following ways:

1. They use annual (fiscal year) data derived from the unified budget series, instead of quarterly data derived from the NIA series.
2. Their sample runs from 1960 through 1984, instead of 1948–1986.
3. They assume that the current surplus net of interest (s_t) is not Granger caused by debt (k_t).
4. As a consequence of (3), they do not formally test the restrictions implied by equation (2.4). Instead they present evidence that expected changes in future surpluses account for a substantial amount of the variation in real debt. In particular, they calculate the squared correlation between the real and implied debt series to be 0.53.

Of the differences listed above, item (3) represents the most serious distinction between the two studies. In assuming that debt does not Granger cause surpluses, HF's approach implies stochastic singularity of $\{(k_{t+1}, s_t)\}$ and eliminates the possibility of performing tests of cross equation restrictions such as those reported in Table 3. This assumption is also strongly rejected by the data: standard tests (not reported here) of causality from debt to surpluses reject noncausality at the 1% significance level. In light of these considerations, the assumption of noncausality does not appear to be justified by the data.

To provide greater comparibility between my results and those in the HF study, the application of the *levels* models tested in Table 3 was modified to more closely resemble that of the HF paper. The data series were modified by annualizing the quarterly data (averaging debt and summing deficits) and restricting the data set to the years 1960-1984. Following HF, a constant *ex ante* real rate of 1.12 percent was assumed, and a constant term and 3 lags were included in the VAR equations. However, Granger noncausality of surpluses by debt was not assumed due to considerations mentioned above. The estimation results for this modified data set are displayed in Table 4. This table show that the results obtainable with the annualized data set are comparable to those obtained in the HF study, in the sense that the debt series implied by the model is highly correlated with the actual debt series. In fact, by dropping the unrealistic assumption that the surplus net of interest is not caused by debt, sample correlations above .9 can be easily obtained. On the other hand, the cross equation restrictions implied by the model are still strongly rejected by the modified data. These results do not qualitatively change when the sample period is expanded to the full data set (1948–86), or restricted to the years before the enactment of the Reagan tax cut (1948–80).

The fact that the model appears to be so consistently and strongly rejected by the postwar U.S. data led me to experiment with a number of different real interest rates and data subsamples, in order to see whether the model represents a reasonable approximation for some subperiod of the postwar data set. The best fit was obtained by specifying the real rate to be very close to zero and restricting the (annual) data set to the pre-oil shock period of 1948–73. For this experiment I was able to obtain $\chi^2(7)$ test statistics of approximately 22. Though this is still highly significant, applying the Schwarz correction for degrees of freedom yields values for the Schwarz criterion that are only slightly unfavorable for the restricted model.

4. Conclusion

The results of Section 3 suggest that the joint hypothesis of a one period ahead, constant *ex ante* real rate version of the government budget constraint and expected net present value budget balance is not a particularly realistic abstraction for the postwar U.S. data. Statistical rejection of the restrictions implied by this hypothesis is robust to assumptions about stationarity of the data, values of the real rate, and the choice of time unit for the model. The model is also rejected for various subsamples of the postwar data set, though the degree of rejection is substantially increased when data from the 1980s is included in the estimation period, and decreased somewhat for very low real rates when only pre-oil shock data is included.

Versions of the government budget constraint that assume a constant *ex ante* real rate are sometimes incorporated in rational expectations macroeconomic models. The results presented here suggest that models incorporating this type of constraint will fail to capture some of the dynamic features of the fiscal data for the U.S. If these data are to be consistent with the idea of expected present value budget balance, then more complex models of the budget balance relationship will have to be formulated. The addition of features such as time varying real rates and a term structure of government debt poses interesting challenges for future research in this area.

Table 1

Tests for Stationarity of Real Debt and Deficits

$$\Delta x_t = \alpha_0 + \alpha_1 \Delta x_{t-1} + \alpha_2 x_{t-1} + \varepsilon_t$$

Series (x_t)

	Debt k_{t+1}	Deficits s_t
Intercept	1.19	−.0403
	(.420)	(−.109)
Δx_{t-1}	.868	.00800
	(20.4)	(.993)
x_{t-1}	−.00104	−.133
	(−.262)	(−3.34)
Q(36)	33.5	35.4

Sample is 1948Q–1986Q4. T-Statistics are in parentheses. To reject the null of a unit root in favor of stationarity, the Dickey-fuller test requires the t statistic on x_{t-1} to be less than -1.95 at the five percent level. Q is the Ljung-Box statistic, distributed $\chi^2(36)$ under the null of white noise residuals.

Table 2

Posterior Distribution for Modulus of the Largest AR Root Λ

$$x_t = \alpha_0 + \sum_{i=1}^{6} \alpha_i x_{t-1} + \varepsilon$$

Series (x_t)	Debt k_{t+1}	Deficits s_t
Posterior Mean	.9748	.8654
Posterior St. Deviation	.02602	.05748
$Pr(\Lambda < 1)$	.8411	.9966
App. Log Odds in favor of a unit root*	.7978	−.7873

Sample is 1949Q3–1986Q4. Calculations are based on 10,000 Monte Carlo replications. See DeJong and Whiteman (1989a) for a detailed description of the Monte Carlo technique.

*Approximate log posterior odds ratio of a unit root versus a stationary alternative. Following suggestion of Sims (1988), the log of this ratio is approximated as −log(posterior variance of Λ)−6.5, which assumes 4 to 1 prior odds in favor of the stationary alternative.

Table 3

Likelihood Ratio Tests of (2.2) and (3.3)

Assumed Constant *Ex Ante* Real Rate (Annualized)

	Lags in VAR	$r = .2\%\%$	$r = 1\%$	$r = 2\%$
Levels Model				
	4	154	144	151
	8	165	156	164
Differences Model				
	4			
	8	129	129	129

Note: Test statistics are distributed $\chi^2(2p+1)$ under the null, where p is the number of lags in the VAR model. Original sample is 1948Q1–1986Q4.

Table 4

Tests of Expected NPV Using Annual Data

Sample Period	1960–84	1948–86	1948–80
Likelihood Ratio $[\chi^2(7)]$	34.6	76.4	40.8
r^2 for Actual vs. Implied Debt Series	.901	.969	.980

Note: tests assume a constant *ex ante* real interest rate of 1.12 percent.

Notes

1. Other related papers are Wilcox (1989) and Shim (1984).

2. Inspection of equation (2.1) reveals that (2.1) is a weaker restriction than equation (2.1) of HSR. Yet a great deal of confusion exists as to whether it is necessary to assume constant real rates to obtain (2.2). Clearly, only a constant *ex ante* real rate need be assumed, since (2.2) follows from (2.1).

3. By linearizing the restrictions implied by (2.4), restrictions are in effect being imposed on the AR rather than the MA representation of the joint process for debt and deficits. Alternatively, imposing restrictions in this fashion amounts to imposing restrictions using the government budget constraint (2.1) instead of the present value relation (2.4). Also, the likelihood ratio tests of these restrictions depend crucially upon the stationarity of the debt process assumed in the derivation of (2.4). It should also be noted that the values of the likelihood ratio test statistics are invariant to the (invertible) transformations used to obtain these linear restrictions.

7

Faster Methods for Solving Continuous Time Recursive Linear Models of Dynamic Economies

by Lars Peter HANSEN, John HEATON and Thomas J. SARGENT

Introduction

This paper describes calculations that are useful for computing equilibria of recursive linear models of general competitive equilibrium in continuous time. The calculations are designed to make it possible to implement continuous time reformulations of a class of models whose discrete time versions are analyzed by Hansen and Sargent (1990). As in Hansen and Sargent's discrete time setting, the basic idea is to formulate a fictitious social planning problem whose solution equals the allocation that would be associated with the competitive equilibrium of a decentralized version of the economy. The social planning problem has a quadratic objective function and linear constraints. Some of the constraints are represented as non-autonomous differential equations, which means that we permit the endowment and preference shocks to be relatively general functions of time. In particular, the preference and endowment shocks need not be restricted to be themselves the outputs of a system of differentiable equations (this is what we mean by saying that the constraints may take the form of *nonautonomous* differential equations). Permitting the preference and endowment shocks to have this general structure is designed eventually to facilitate a number of applications that we have in mind. The intention is to formulate the social planning problem with ample generality by allowing room to include a variety of setups with potentially large numbers of capital stocks and informational state variables. As in Hansen and Sargent (1990), forcing the planning problem into the form of a linear-quadratic optimization problem yields substantial computational benefits: the equilibrium allocations can be computed by solving that optimization problem (which is often called an *optimal linear regulator problem*), and the Arrow-Debreu prices that support those allocations as competitive equilibria can be computed from information contained in the value function for that optimization problem.

There are standard methods available for solving a class of problems similar to ours known as optimal linear regulator problems (see Kwakernaak and Sivan 1972 and Anderson 1978). However, these algorithms require that we can represent the constraints as systems of autonomous differential equations. This paper shows how those standard methods can be adapted to handle problems in which some of the constraints take the form of nonautonomous differential equations. We proceed recursively, first solving a reduced system whose constraints are represented as an autonomous system of differential equations. In solving this reduced problem, we use fast algorithms to compute the *feedback* part of the solution of the original problem. Second, we compute an additional part of the solution, called the *feedforward* part, which incorporates the effect of the nonautonomous parts of the differential equations describing the constraints. Fast algorithms for implementing both parts of these calculations are described in this paper.

This paper can be viewed as an extended technical prolegomenon to future work that will describe a continuous-time version of Hansen-Sargent (1990). In effect, we describe continuous time versions of an array of computational tricks that have discrete time analogues that are studied by Hansen and Sargent (1990). We concentrate our attention on solving the social planning problem, and do not lay out the connection of the social planning problem to a continuous time version of a competitive equilibrium. However, it is important to note that the connection to a competitive equilibrium is there, and that it can be spelled out by mimicking in continuous time versions of the arguments that Hansen and Sargent utilize in discrete time.

This paper studies a nonstochastic version of the model. However, just as in Hansen-Sargent (1990), once the solution of the planning problem under certainty is obtained, it is a simple matter to compute the equilibrium of corresponding stochastic economies. This is a standard feature of economies whose planning problems are optimal linear regulators. In the next paper in this volume, a set of continuous time prediction formulas that have been devised to compute the *feedforward* parts of a stochastic version of our models. As emphasized by Lucas and Sargent (1981) a convenient feature of linear quadratic models is the manner in which their solution *separates* into a control part and a predition part. This paper and the next one correspond to this separation.

This paper is organized as follows. Section 1 sets forth the social planning or optimal resource allocation problem. The planning problem assumes the form of an optimal linear regulation problem. Section 2 defines several linear operators that are useful in deriving and representing a solution. Section 3 uses some of these operators to represent the constraints impinging on the planning problem in a way that facilitates computing a solution. Section 4 solves our particular optimal linear regulator problem in a way that exploits its special structure. Section 5 describes how a *matrix sign* algorithm can be applied to simplify and speed aspects of the computations. Sections 6 and 7 are devoted to studying two special cases of our model. A class of continuous time models of costly adjustment of capital are described in Section 6, while Section 7 studies a version of Heaton's (1989) model of consumer durables.

1. Optimal Resource Allocation Problem

We are interested in computing solutions to an optimal resource allocation problem that is the continuous-time counterpart to a discrete-time problem studied extensively by Hansen and Sargent (1990). The solution to this optimization problem can be decentralized and interpreted as the time path for the equilibrium quantities of an intertemporal competitive equilibrium.

Consider the following general setup. A household has time separable preferences defined over an n_s -dimensional vector $s(t)$ of *services* at time t. The preferences are given by:

$$(1.1) \qquad -(1/2)\int_0^\infty \exp(-\rho t)\ [s(t)-b(t)]'\ [s(t)-b(t)]dt$$

where $b(t)$ is an exogenous n_s -dimensional vector function of time and $\rho > 0$ is the subjective rate of discount. Services at time t are generated according to:

$$(1.2) \qquad s(t) = \Lambda h(t) + \Pi c(t)$$

where $h(t)$ is a n_h -dimensional vector of *household capital* stocks at time t, $c(t)$ is an n_c -dimensional vector of *consumption goods* at time t, Λ is an $n_s \times n_h$ matrix and Π is an $n_s \times n_c$ matrix.

The household capital stock evolves according to the following system of linear differential equations:

$$(1.3) \qquad Dh(t) = \Delta_h\, h(t) + \Theta_h\, c(t), \quad \text{for} \quad t \geq 0,\ h(0) = \mu_h\ ,$$

where Δ_h is an $n_h \times n_h$ matrix, Θ_h is an $n_h \times n_c$ matrix and μ_h is an $n_h \times 1$ vector of initial conditions. The matrix Δ_h governs the depreciation of the household capital stock.

Taken together, (1.2) and (1.3) induce intertemporal nonseparabilities into the indirect preference ordering for consumption. Among other things, the matrices Λ and Π determine the extent to which the consumption goods are substitutes or complements over time.

To allow the transfer of output over time, there is an n_k vector $k(t)$ consisting of *physical capital* stocks at time t. Capital is assumed to evolve over time according to:

$$Dk(t) = \Delta_k\, k(t) + \Theta_k\, i(t), \quad \text{for} \quad t \geq 0, \; k(0) = \mu_k \tag{1.4}$$

where $i(t)$ is an n_i vector of *investment goods* at time t, Δ_k is an $n_k \times n_k$ matrix, Θ_k is an $n_k \times n_i$ matrix and μ_k is an $n_k \times 1$ vector of initial conditions. The matrix Δ_k determines the depreciation of the physical capital stocks.

Output is produced using the current capital stock and an endowment $f(t)$. This output is then divided between consumption and investment goods according to:

$$\Phi_c\, c(t) + \Phi_i\, i(t) = \Gamma\, k(t) + f(t) \tag{1.5}$$

where Φ_c is an n_c -dimensional nonsingular matrix, Φ_i is an $n_c \times n_i$ matrix and Γ is an $n_c \times n_k$ matrix.

The functions $f(t)$ and $b(t)$ are driven by an n_z -dimensional vector of forcing functions of time denoted $\hat{z}(t)$:

$$f(t) = \Xi_f\, \hat{z}(t) \quad \text{and} \quad b(t) = \Xi_b\, \hat{z}(t) \tag{1.6}$$

where Ξ_f is an n_c by n_z matrix and Ξ_b is an n_s by n_z matrix.

Suppose now that we define a composite state vector $\hat{x}(t)' \equiv [h(t)'\ k(t)']$ and a control vector $\hat{u}(t) \equiv i(t)$. Using the resource constraint (1.5) and the differential equation systems (1.3) and (1.4), we get the following system of differential equations governing $\hat{x}(t)$:

$$D\hat{x}(t) = \hat{A}\hat{x}(t) + B_u\, \hat{u}(t) + B_z\hat{z}(t) \qquad \text{subject to} \quad \hat{x}(0) = \mu \tag{1.7}$$

where $\hat{A} \equiv \begin{bmatrix} \Delta_h & \Theta_h(\Phi_c)^{-1}\Gamma \\ 0 & \Delta_k \end{bmatrix}$, $B_u \equiv \begin{bmatrix} -\Theta_h(\Phi_c)^{-1}\Phi_i \\ \Theta_k \end{bmatrix}$, $B_z \equiv \begin{bmatrix} \Theta_h(\Phi_c)^{-1}\Xi_f \\ 0 \end{bmatrix}$ and $\mu' \equiv (\mu_h', \mu_k')$. Notice that

$$s(t) - b(t) = \Sigma_u\, \hat{u}(t) + \Sigma_x\hat{x}(t) + \Sigma_z\hat{z}(t) \tag{1.8}$$

where $\Sigma_u \equiv -\Pi(\Phi_c)^{-1}\Phi_i$, $\Sigma_x \equiv [\Lambda \ ; \ \Pi(\Phi_c)^{-1}\Gamma]$ and $\Sigma_z \equiv \Pi(\Phi_c)^{-1} \Xi_f - \Xi_b$. We can now write the objective function (1.1) in terms of $\hat{u}(t)$, $\hat{x}(t)$ and $\hat{z}(t)$:

$$(1.9) \qquad -(1/2)\int_0^\infty \exp(-\rho t)\ [\hat{u}(t)'\ \hat{x}(t)'\ \hat{z}(t)']\,\Omega \begin{bmatrix} \hat{u}(t) \\ \hat{x}(t) \\ \hat{z}(t) \end{bmatrix} dt$$

where $\Omega \equiv \begin{bmatrix} \Omega_{uu} & \Omega_{ux} & \Omega_{uz} \\ \Omega'_{ux} & \Omega_{xx} & \Omega_{xz} \\ \Omega'_{uz} & \Omega'_{xz} & \Omega_{zz} \end{bmatrix}$. The partitions of Ω are given by $\Omega_{uu} = \Sigma'_u\Sigma_u$, $\Omega_{ux} = \Sigma'_u\Sigma_x$, and so on. Notice that Ω is positive semidefinite.

We assume that the capital stocks and forcing functions satisfy:

$$(1.10) \qquad \begin{array}{l} \int_0^\infty \exp(-\rho t) \mid h(t) \mid^2 dt < \infty, \ \int_0^\infty \exp(-\rho t) \mid k(t) \mid^2 dt < \infty \\ \int_0^\infty \exp(-\rho t) \mid \hat{z}(t) \mid^2 dt < \infty \, . \end{array}$$

These constraints limit growth in the respective functions of time. In the case of the capital stocks, these constraints are used instead of nonnegativity constraints because they are easier to impose. The plausibility of the resulting solutions can be checked in practice by solving the problem numerically subject to these constraints and checking that the resulting time paths for the capital stocks are nonnegative.

Prior to solving the model, it is convenient to transform the optimization problem to remove the discounting. This is done as follows. Let $\epsilon = \rho/2$ and define

$$(1.11) \qquad \begin{array}{rl} & x(t) \equiv \exp(-\epsilon t)\hat{x}(t), \quad z(t) \equiv \exp(-\epsilon t)\hat{z}(t) \\ \text{and} & u(t) \equiv \exp(-\epsilon t)\hat{u}(t) \, . \end{array}$$

In terms of the transformed vector $x(t)$, the restrictions in (1.10) imply

$$(1.10') \qquad \int_0^\infty \mid x(t) \mid^2 dt < +\infty.$$

Notice that

$$(1.12) \qquad Dx(t) = -\epsilon x(t) + \exp(-\epsilon t) D\hat{x}(t) \, .$$

As a consequence $x(t)$ satisfies the following system of differential equations:

$$(1.13) \qquad Dx(t) = Ax(t) + B_u u(t) + B_z z(t) \, , \quad \text{for } \ t \geq 0, \ x(0) = \mu$$

where $A = \hat{A} - \epsilon I$. Also the objective function (1.9) can be written as:

$$(1.14) \qquad -(1/2)\int_0^\infty [u(t)'\ x(t)'\ z(t)']\ \Omega \begin{bmatrix} u(t) \\ x(t) \\ z(t) \end{bmatrix} dt\ .$$

Our problem now is to maximize (1.14) subject to (1.10′) and (1.13) by choosing $\{u(t)\}_{t=0}^{\infty}$.

2. Important Linear Operators

Let C denote the space of complex numbers and R the space of real numbers. Also, let L_1 denote the collection of Borel measurable functions x mapping R into the space of complex numbers C such that

$$(2.1) \qquad \int_R |\ x(t)\ |\ dt < \infty$$

and let L_2 denote the collection of Borel measurable functions x such that

$$(2.2) \qquad \int_R |\ x(t)\ |^2\ dt < \infty\ .$$

Let L_1^n and L_2^n be the spaces of all n-dimensional vectors of functions with components in L_1 and L_2 respectively. In light of (1.10′), the vector of functions x given in (1.11) can be restricted to be elements of L_2^n where $n = n_h + n_k$.

It is convenient to analyze four operators that map L_2^n into itself using Fourier methods. Among other things, these operators will be useful in characterizing a particular solution to the differential equation system (1.13). Our strategy for constructing these operators is as follows. First, we display the operators evaluated at functions in $L_1^n \cap L_2^n$. Second, we verify that the operators map $L_1^n \cap L_2^n$ into L_2^n. Consequently, we can compute Fourier transforms of the resulting functions. The Fourier transforms of the operators, i.e., the composition of the Fourier transform operator with the original operators, have simple extensions to all of L_2^n. The original operators can then be extended to all of L_2^n by taking inverse Fourier transforms. These steps will be made clear in our examples.

The first operator is an *aggregation over time* operator. It aggregates x over γ time units. It is defined on $L_1^n \cap L_2^n$ via

$$(2.3) \qquad U_\gamma(x)\,(t) = \int_o^\gamma x(t+\tau)d\tau$$

where x is in $L_1^n \cap L_2^n$. It is straightforward to show that $U_\gamma(x)$ is in L_1^n. For any function x in L_1^n we define the Fourier transform operator T evaluated at x to be:

$$T(x)(\theta) = \int_R \exp(-i\theta t)\, x(t)dt \tag{2.4}$$

We evaluate the Fourier transform of $U_\gamma(x)$ by changing orders of integration. This gives:

$$\begin{aligned} T[U_\gamma(x)](\theta) &= \int_R \exp(-i\theta t)\, [\int_0^\gamma x(t+\tau)d\tau]dt \\ &= \int_0^\gamma \exp(i\theta\tau)\, [\int_R \exp[-i\theta(t+\tau)]x(t+\tau)dt]d\tau \\ &= \{[\exp(i\theta\gamma) - 1]/(i\theta)\}T(x)(\theta)\,. \end{aligned} \tag{2.5}$$

The function ϕ, where $\phi(\theta) \equiv [\exp(i\theta\gamma) - 1]/(i\theta)$, is continuous in θ on R including zero. In addition, $\phi(\theta)$ goes to zero as $|\theta|$ goes to infinity. Therefore, ϕ is bounded in θ. It follows from a multidimensional analog to the Parseval formula that $U_\gamma(x)$ is in $L_1^n \cap L_2^n$ (e.g. see Rudin 1974, Chapter 9).

Notice that the operator obtained by multiplying x in L_2^n by ϕ maps L_2^n into L_2^n. We extend the construction of U_γ to all of L_2^n by letting $U_\gamma(x)$ be the element of L_2^n that has $\phi T(x)$ as its Fourier transform. In Appendix A we establish that this construction of U_γ is consistent with (2.3) and (2.5) for all x in L_2^n.

The second operator we consider is the *shift* operator. For any x in $L_1^n \cap L_2^n$ we define

$$S_\gamma(x)(t) = x(t+\gamma)\,. \tag{2.6}$$

Taking Fourier transforms of both sides of (2.6) gives

$$\begin{aligned} T[S_\gamma(x)](\theta) &= \int_R \exp(-i\theta t)x(t+\gamma)dt \\ &= \exp(i\theta\gamma) \int_R \exp[-i\theta(t+\gamma)]x(t+\gamma)dt \\ &= \exp(i\theta\gamma)T(x)(\theta)\,. \end{aligned} \tag{2.7}$$

These calculations can be extended from $L_1^n \cap L_2^n$ to L_2^n just as in the previous case except that now $\phi(\theta) \equiv \exp(i\theta\gamma)$. Note that $|\phi(\theta)| = 1$ for all θ. In Appendix A we show that this extension is compatible with (2.6).

The third operator is a *matrix backward convolution* operator. Again let x be in $L_1^n \cap L_2^n$, and let A be an $n \times n$ matrix of real numbers with eigenvalues that have strictly negative real parts. The *backward* convolution operator is then given by:

$$C(x)(t) = \int_0^\infty \exp(A\tau)x(t-\tau)d\tau \tag{2.8}$$

where

$$\exp(A) = \Sigma_{j=0}^{\infty} A^j/j\ ! \tag{2.9}$$

The components of $C(x)$ are in L_1 since the eigenvalues of A have strictly negative real parts and

$$\begin{aligned} \int_R |C(x)(t)|dt &\leq \int_R \int_0^\infty |\exp(A\tau)|\ |x(t-\tau)|\, d\tau dt \\ &= \int_0^\infty |\exp(At)|\, dt \int_R |x(t)|\, dt\ . \end{aligned} \tag{2.10}$$

Hence formula (2.4) implies that
(2.11)

$$\begin{aligned} T[C(x)](\theta) &= \int_R \exp(-i\theta t)[\int_0^\infty \exp(A\tau)x(t-\tau)d\tau]dt \\ &= \int_0^\infty \exp(-i\theta\tau)\exp(A\tau)\{\int_R \exp[-i\theta(t-\tau)x(t-\tau)dt\}d\tau \\ &= (i\theta I - A)^{-1}T(x)(\theta)\ . \end{aligned}$$

Notice that $i\theta I - A$ is nonsingular for all θ in R because the eigenvalues of A have strictly negative real parts. Consequently,

$$\bar{\delta} \equiv \sup_{\theta \in R} \text{trace}\{(-i\theta I - A')^{-1}(i\theta I - A)^{-1}\} < \infty\ . \tag{2.12}$$

Applying the Cauchy-Schwarz Inequality gives

$$\int_0^\infty |\ T[C(x)(\theta)]\ |^2\, d\theta \leq \bar{\delta} \int_0^\infty |\ T(x)(\theta)\ |^2\, d\theta\ . \tag{2.13}$$

Thus $T[C(x)]$ has components in L_2^n, and we extend C to all of L_2^n so that (2.11) is satisfied. In Appendix A we show that this extension is compatible with (2.8).

Using the same logic, we can define a *forward convolution* operator C' such that for any x in L_2^n,

$$T[C'(x)](\theta) = (-i\theta I - A')^{-1}T(x)(\theta)\ . \tag{2.14}$$

When x is also in L_1^n, the transform operator has a time domain representation:

$$C'(x)(t) = \int_0^\infty \exp(A'\tau)x(t+\tau)d\tau \;. \tag{2.15}$$

Thus the transform operator is a *forward* convolution operator. The operator $C'(x)$ can also be extended to all of L_2^n in a manner that preserves (2.15).

3. Linear Constraints

In Section 1, we described an optimization problem that is subject to a linear constraint that can be represented as a system of differential equations:

$$Dx(t) = Ax(t) + Bv(t) \;\text{ for }\; t \geq 0, \; x(0) = \mu \tag{3.1}$$

where $B = [B_u \; B_z]$ and $v(t)' = [u(t)' \; z(t)']$. We refer to equation (3.1) as the *state equation*. In this section we describe how to solve the state equation and obtain a frequency domain representation of the constraints. As in Section 2, we restrict the eigenvalues of A:

Assumption 3.1: The eigenvalues of the matrix A have real parts that are strictly negative.

If the matrices Δ_h and Δ_k discussed in Section 1 all have eigenvalues with real parts that are strictly less than ϵ, then the model of Section 1 will satisfy Assumption 3.1.

We restrict v to be in L_2^N where $N = n_i + n_z$. For convenience, we set $v(t)$ to zero for strictly negative values of t. The space of all elements of L_2^N that are zero for negative t will be denoted L_+^N. We now consider solving (3.1) for x in terms of v in L_+^N. We deduce a particular solution to (3.1) that applies for all time periods but the solution will not necessarily satisfy $x(0) = \mu$. The particular solution will be denoted x^p.

For equation (3.1) to be satisfied for all t, x^p must satisfy:

$$x^p(t+\gamma) - x^p(t) = A\int_t^{t+\gamma} x^p(s)ds + B\int_t^{t+\gamma} v(s)ds \tag{3.2}$$

for all t and all strictly positive values of γ. Expressing (3.2) in operator notation gives

$$S_\gamma(x^p) - x^p = AU_\gamma(x^p) + BU_\gamma(v) \;. \tag{3.3}$$

Taking Fourier transforms gives

$$\begin{aligned}(3.4)\quad &[\exp(i\theta\tau) - 1]T(x^p)(\theta) \\ &= \{[\exp(i\theta\tau) - 1]/(i\theta)\}\,[AT(x^p)(\theta) + BT(v)(\theta)]\,.\end{aligned}$$

Solving (3.4) for $T(x^p)$ gives

$$(3.5)\qquad T(x^p)(\theta) = (i\theta I - A)^{-1}BT(v)(\theta)\,.$$

From the analysis in Section 2, we know that $x^p = C(Bv)$, or

$$(3.6)\qquad x^p(t) = \int_0^\infty \exp(A\tau)Bv(t-\tau)d\tau\;,$$

[see (2.11)]. Equations (3.1) and (3.5) suggest a convenient notation for the convolution operator that is analogous to the notation used for lag operators in discrete time. The backward convolution operator applied to an n-dimensional vector of functions, x, will be written as:

$$(3.7)\qquad C(x)(t) \equiv (DI - A)^{-1}x(t)\,.$$

where we have replaced $i\theta$ by D.

Using this new notation, we represent x^p as:

$$(3.8)\qquad x^p(t) = (DI - A)^{-1}Bv(t)\,.$$

Since $v(t)$ is zero for strictly negative values of t, it follows that x^p is in L_+^n and that for $t \geq 0$

$$(3.9)\qquad x^p(t) = \int_0^t \exp(A\tau)Bv(t-\tau)d\tau\,.$$

The process $\{x^p(t) : -\infty < t < +\infty\}$ given by (3.9) does not, in general, satisfy the initial condition because (3.9) implies that $x^p(0)$ is zero. The homogeneous differential equation

$$(3.10)\qquad Dx^h(t) = Ax^h(t) \;\text{ for }\; t \geq 0,\; x(0) = \mu$$

has as its solution

$$(3.11)\qquad x^h(t) = \exp(At)\mu\,.$$

This can be verified by differentiating the power series expansion for $\exp(At)$ with respect to t.

Adding together x^p to x^h gives a solution to the original differential equation (3.1). Hence

$$(3.12) \qquad x(t) = \exp(At)\mu + \int_0^t \exp(A\tau)Bv(t-\tau)d\tau \quad \text{for} \quad t \geq 0 \ .$$

Equality (3.12) will be taken as the constraint for the optimization problem we study in Section 4.

There is a differential equation system that is closely related to equation system (3.1). Let x and λ be processes in L_2^n such that

$$(3.13) \qquad Dx(t) = -A'x(t) - \lambda(t) \ .$$

Notice that the eigenvalues of $-A'$ have real parts that are strictly positive. The function λ will be a Lagrange multiplier in our analysis in Section 4.

We proceed as before by using Fourier transform methods to solve (3.13) for x in terms of λ. Expressed in terms of transforms, this gives

$$(3.14) \qquad T(x)(\theta) = (-i\theta I - A')^{-1}T(\lambda)(\theta) \ .$$

It follows from the analysis in Section 2 that:

$$(3.15) \qquad x(t) = C'(\lambda)(t) = \int_0^\infty \exp(A'\tau)\lambda(t+\tau)d\tau \ .$$

Just as in the case of the backward convolution operator, (3.13) and (3.15) suggest a convenient notation for the forward convolution operator:

$$(3.16) \qquad C'(\lambda)(t) \equiv (-DI - A')^{-1}\lambda(t) \ .$$

Hence we write a solution to equation (3.13) as:

$$(3.17) \qquad x(t) = (-DI - A')^{-1}\lambda(t) \ .$$

When (3.15) is only required to hold for $t \geq 0$, one may consider adding a term $\exp(-A't)\eta$ to solution (3.15) for x. In this case, however, any nonzero value of η results in a function x that is not in L_2^n no matter how this function is extended to the entire real line. Therefore, (3.15) gives the entire class of solutions in L_2^n.

4. Optimal Linear Regulator Problem

In this section we solve the following optimal linear regulator *(OLR)* problem.

<u>*OLR* Problem:</u> $$\max_{u\in L_+^N i, x\in L_+^n} -(1/2)\int_0^\infty [u(t)'\ x(t)'\ z(t)']\Omega \begin{bmatrix} u(t) \\ x(t) \\ z(t) \end{bmatrix} dt$$

subject to $x(t) = \exp(At)\mu + \int_0^t \exp(A\tau)\,[B_u u(t-\tau) + B_z z(t-\tau)]dt$ for $t \geq 0$.

Our formulation of the *OLR* problem differs from the standard formulation found, say, in Kwakernaak and Sivan (1972), in two ways. First, we allow z to be any element of $L_+^{n_z}$, not necessarily the solution to a linear differential equation system. Second, for systems in which it is not optimal to stabilize the state (i.e., for x to be in L_+^n), we impose stability as an extra constraint.

Without further restrictions, the *OLR* does not always have a solution. The following restriction on A, B_u and Ω guarantees the existence of a solution.

<u>Assumption 4.1:</u> The matrix $\Psi_{uu}(i\theta)$ is Hermitian, and there exists a strictly positive real number δ such that $\Psi_{uu}(i\theta) > \delta I$ for all θ in R, where

$$\Psi_{uu}(\zeta) \equiv [I\,;\ B_u'(-\zeta I - A')^{-1}]\Omega_{11} \begin{bmatrix} I \\ (\zeta I - A)^{-1}B_u \end{bmatrix} \tag{4.1}$$

and

$$\Omega_{11} = \begin{bmatrix} \Omega_{uu} & \Omega_{ux} \\ \Omega_{xu} & \Omega_{xx} \end{bmatrix}. \tag{4.2}$$

Note that for each θ in R, $\Psi_{uu}(i\theta)$ is a positive semidefinite matrix. Assumption 4.1 puts a lower bound on the eigenvalues of $\Psi(i\theta)$ for all θ. The inequality restriction must apply to the limit as θ goes to infinity as well. Evaluating this limit, we see that Assumption 4.1 implies that

$$\lim_{|\theta|\to\infty} \Psi_{uu}(i\theta) = \Omega_{uu} \geq \delta I \tag{4.3}$$

Therefore Ω_{uu} is nonsingular. In Appendix B we verify that when Assumptions 3.1 and 4.1 are satisfied, there exists a solution to the *OLR* problem for any initial condition μ.

We solve the *OLR* problem using Lagrange multipliers. Since L_+^n is a Hilbert space, continuous linear functionals on this space are representable as inner products with elements of L_+^n. Consequently, the Lagrange multiplier λ associated with constraint (3.12) is an element of L_+^n and the Lagrangean is given by:

(4.4)

$$\mathcal{L} \equiv - \int_0^\infty \left\{ 1/2[u(t)'\, x(t)'\, z(t)']\Omega \begin{bmatrix} u(t) \\ x(t) \\ z(t) \end{bmatrix} - \lambda(t) \cdot [x(t) - x^p(t) - x^h(t)] \right\} dt$$

It follows from the Parseval formula that

(4.5)

$$\begin{aligned}
\int_0^\infty \lambda(t) \cdot x^p(t)\, dt &= (1/2\pi) \int_R T(\lambda)(\theta)^{*\prime} T(x^p)(\theta)\, d\theta \\
&= (1/2\pi) \int_R T(\lambda)(\theta)^{*\prime} (i\theta I - A)^{-1} BT(v)(\theta)\, d\theta \\
&= (1/2\pi) \int_R \{(-i\theta I - A')^{-1} T(\lambda)(\theta)\}^{*\prime} [B_u T(u)(\theta) \\
&\qquad + B_z T(z)(\theta)] d\theta \\
&= \int_0^\infty \{[B_u' (-DI - A')^{-1} \lambda(t)\} \cdot u(t) dt \\
&\qquad + \int_0^\infty \{[B_z' (-DI - A')^{-1} \lambda(t)\} \cdot z(t) dt\, .
\end{aligned}$$

The first-order conditions for Problem 1 are obtained by differentiating $\mathcal{L}$ with respect to u and x, giving

$$\Omega_{uu}\, u_s(t) + \Omega_{ux}\, x_s(t) + \Omega_{uz}\, z(t) + B_u' (-DI - A')^{-1} \lambda_s(t) = 0 \tag{4.6}$$

$$\Omega_{xu}\, u_s(t) + \Omega_{xx}\, x_s(t) + \Omega_{xz}\, z(t) = \lambda_s(t)\, . \tag{4.7}$$

The subscripts s are included on x, u, and λ to denote the solution.

We solve equation (4.6) for $u_s(t)$ giving

$$\begin{aligned} u_s(t) = &- (\Omega_{uu})^{-1} B_u' (-DI - A')^{-1} \lambda_s(t) - (\Omega_{uu})^{-1} \Omega_{ux}\, x_s(t) \\ &- (\Omega_{uu})^{-1} \Omega_{uz}\, z(t)\, . \end{aligned} \tag{4.8}$$

Substituting (4.8) into (4.7) yields

$$\begin{aligned} \lambda_s(t) = &- \Omega_{xu} (\Omega_{uu})^{-1} B_u' (-DI - A')^{-1} \lambda_s(t) \\ &+ [\Omega_{xx} - \Omega_{xu} (\Omega_{uu})^{-1} \Omega_{ux}] x_s(t) \\ &+ [\Omega_{xz} - \Omega_{xu} (\Omega_{uu})^{-1} \Omega_{uz}] z(t)\, . \end{aligned} \tag{4.9}$$

For each time $t \geq 0$, define a co-state vector $x_c(t)$ via

$$x_c(t) \equiv (-DI - A')^{-1} \lambda_s(t) . \tag{4.10}$$

Then the analysis in Section 3 implies that the co-state equation

$$Dx_c(t) = -A'x_c(t) - \lambda_s(t) \tag{4.11}$$

is satisfied [see (3.13) - (3.17)]. Substituting (4.9) and (4.10) into (4.11) gives

(4.12)
$$\begin{aligned} Dx_c(t) = {} & [\Omega_{xu} (\Omega_{uu})^{-1} \Omega_{ux} - \Omega_{xx}]x_s(t) - [A - B_u(\Omega_{uu})^{-1}\Omega_{ux}]'x_c(t) \\ & + [\Omega_{xu} (\Omega_{uu})^{-1}\Omega_{uz} - \Omega_{xz}]z(t) . \end{aligned}$$

Substituting (4.8) and (4.10) into the state equation (1.13) yields

$$\begin{aligned} Dx_s(t) = {} & [A - B_u(\Omega_{uu})^{-1}\Omega_{ux}]x_s(t) - B_u(\Omega_{uu})^{-1}B_u'x_c(t) \\ & + [B_z - B_u(\Omega_{uu})^{-1}\Omega_{uz}]z(t) . \end{aligned} \tag{4.13}$$

To solve for $x_s(t)$ and $x_c(t)$, we combine equations (4.12) and (4.13) into a single system:

$$\begin{bmatrix} Dx_s(t) \\ Dx_c(t) \end{bmatrix} = H \begin{bmatrix} x_s(t) \\ x_c(t) \end{bmatrix} + Kz(t) . \tag{4.14}$$

where

$$H = \begin{bmatrix} H_{11} & H_{12} \\ H_{21} & H_{22} \end{bmatrix}, \quad K = \begin{bmatrix} K_1 \\ K_2 \end{bmatrix} , \tag{4.15}$$

$H_{11} \equiv A - B_u(\Omega_{uu})^{-1}\Omega_{ux}$, $H_{12} \equiv -B_u(\Omega_{uu})^{-1} B_u'$, $H_{21} \equiv -\Omega_{xx} + \Omega_{xu}(\Omega_{uu})^{-1}\Omega_{ux}$ and $H_{22} \equiv -[A - B_u(\Omega_{uu})^{-1}\Omega_{ux}]'$, $K_1 \equiv B_z - B_u(\Omega_{uu})^{-1}\Omega_{uz}$ and $K_2 \equiv \Omega_{xu}(\Omega_{uu})^{-1}\Omega_{uz} - \Omega_{xz}$. The matrix H is referred to as a Hamiltonian matrix and satisfies two properties that are important for our purposes. First $H_{22} = -H_{11}'$; and second, H_{12} and H_{21} are symmetric.

Consider first the case in which $z(t)$ is zero for all t. In this case we solve the homogeneous differential equation system:

$$\begin{bmatrix} Dx_s(t) \\ Dx_c(t) \end{bmatrix} = H \begin{bmatrix} x_s(t) \\ x_c(t) \end{bmatrix} \tag{4.16}$$

We follow Vaughan (1969) and solve (4.16) by taking the Jordan decomposition of H:

$$H = EJ(E)^{-1} \tag{4.17}$$

where J is a matrix containing the eigenvalues of H on the diagonal. Some of the entries of J immediately to the right of the eigenvalues that occur more than once are one, and the remaining entries of J are zero. For notational convenience, we place the eigenvalues with strictly negative real parts in the upper left block of J.

The eigenvalues of H occur in symmetric pairs *vis-a-vis* the imaginary axis of the complex plane. To see this, let ζ be any eigenvalue of H, and let $[e_1', e_2']'$ be the corresponding column eigenvector of H. Then $[e_2', -e_1']$ is a row eigenvector of H with eigenvalue $-\zeta$. As a consequence of this symmetry, there can only be at most n eigenvalues of H with strictly negative real parts. We will show, in fact, that there are exactly n such eigenvalues.

It is convenient to transform the equation system (4.14). Define

$$x^+(t) = E^{-1} \begin{bmatrix} x_s(t) \\ x_c(t) \end{bmatrix} \tag{4.18}$$

Then $x^+(t)$ satisfies

$$Dx^+(t) = Jx^+(t) . \tag{4.19}$$

The advantage of working with differential equation system (4.19) is the dynamics are uncoupled according to the distinct eigenvalues.

We partition $x^+(t)' = [x_n^+(t)', x_p^+(t)']$ where the number of entries of $x_n^+(t)$ is equal to the number of eigenvalues of H with strictly negative real parts. Since the remaining eigenvalues have nonnegative real parts, the function x_p^+ will not be stable (have entries in L_+^1) unless $x_p^+(0)$ is zero. Thus we have two initial conditions: $x_s(0) = \mu$ and $x_p^+(0) = 0$.

It follows from (4.18) that

$$E \begin{bmatrix} x_n^+(t) \\ 0 \end{bmatrix} = \begin{bmatrix} x_s(t) \\ x_c(t) \end{bmatrix} . \tag{4.20}$$

Partition E as

$$E = \begin{bmatrix} E_{11} & E_{12} \\ E_{21} & E_{22} \end{bmatrix} . \tag{4.21}$$

Then for $t = 0$, (4.20) can be expressed as

$$E_{11}\, x_n^+(0) = \mu \quad \text{and} \quad E_{21}\, x_n^+(0) = x_c(0) \tag{4.22}$$

Since a solution is known to exist (see Appendix B), for any μ there must exist a vector $x_n^+(0)$ such that (4.22) is satisfied. Consequently, E_{11} must have n columns (H must have n eigenvalues with strictly negative real parts) and E_{11} must be nonsingular. Solving (4.22) for $x_c(0)$ gives

$$x_c(0) = M_x\, \mu\ , \tag{4.23}$$

where $M_x \equiv E_{21}(E_{11})^{-1}$ and solving (4.20) for $x_c(t)$ gives

$$x_c(t) = M_x x_s(t)\ . \tag{4.24}$$

It is of interest to obtain recursive solutions for $Dx_s(t)$ and $Dx_c(t)$. Substituting (4.24) into (4.18) gives

$$\begin{bmatrix} Dx_s(t) \\ Dx_c(t) \end{bmatrix} = \begin{bmatrix} (H_{11} + H_{12}M_x) \\ (H_{21} - H_{11}'M_x) \end{bmatrix} x_s(t) \tag{4.25}$$

The eigenvalues of $H_{11} + H_{12}M_x$ are the same as the stable eigenvalues of H. It follows from (4.24) that $Dx_c(t) = M_x Dx_s(t)$ and from (4.25) that

$$-M_x H_{11} - M_x H_{12} M_x + H_{21} - H_{11}' M_x = 0\ . \tag{4.26}$$

An equivalent statement of (4.26) is

$$[-M_x\ I]\, H \begin{bmatrix} I \\ M_x \end{bmatrix} = 0\ . \tag{4.27}$$

In Appendix C we show that the matrix M_x turns out to be symmetric and positive semidefinite. Furthermore, the time zero value function, or equivalently the optimal value of the criterion function as a function of the initial condition for $x(t)$, is given by $(-1/2)\mu' M_x \mu$.

We now consider the general case in which $z(t)$ is allowed to be different from zero. In this case it is convenient to use a somewhat different transformation than (4.18). Define a matrix M

$$M \equiv \begin{bmatrix} I & 0 \\ -M_x & I \end{bmatrix} \tag{4.28}$$

and transform the composite state-costate function:

$$\begin{bmatrix} x_s(t) \\ w(t) \end{bmatrix} \equiv M \begin{bmatrix} x_s(t) \\ x_c(t) \end{bmatrix} . \tag{4.29}$$

Then

$$\begin{bmatrix} Dx_s(t) \\ Dw(t) \end{bmatrix} = MHM^{-1} \begin{bmatrix} x_s(t) \\ w(t) \end{bmatrix} + MKz(t) \tag{4.30}$$

Since M_x is symmetric and (4.27) is satisfied,

$$MHM^{-1} = \begin{bmatrix} (H_{11} + H_{12}M_x) & H_{12} \\ 0 & -(H_{11} + H_{12}M_x)' \end{bmatrix} \tag{4.31}$$

and

$$MK = \begin{bmatrix} K_1 \\ -M_x K_1 + K_2 \end{bmatrix} . \tag{4.32}$$

It is possible to solve the second block of equations in (4.30) separately because the matrix $MH(M)^{-1}$ is upper triangular. The eigenvalues of the matrix $-(H_{11}+H_{12}M_x)'$ have strictly positive real parts: therefore, the stable solution for $w(t)$ is given by the following forward convolution:

$$\begin{aligned} w(t) &= [-DI - (H_{11} + H_{12}M_x)']^{-1} (K_2 - M_x K_1) z(t) \\ &= \int_0^{\infty} \exp[(H_{11} + H_{12}M_x)'\tau] \, (K_2 - M_x K_1) z(t+\tau) d\tau \end{aligned} \tag{4.33}$$

It follows from (4.30) that

$$Dx_s(t) = (H_{11} + H_{12} M_x) x_s(t) + H_{12} w(t) + K_1 z(t) . \tag{4.34}$$

We refer to $(H_{11} + H_{12} M_x)x_s(t)$ as the *feedback* part and $H_{12} w(t) + K_1 z(t)$ as the *feedforward* part of the decision rule for $Dx_s(t)$. Notice that the transpose of the feedback matrix $(H_{11} + H_{12} M_x)$ enters the exponential in the feedforward integral.

5. Recursive Solution Methods

In this section we describe recursive methods for computing two objects. First we show how to calculate the matrix M_x used in the feedback and feedforward portions of the decision rule for $Dx_s(t)$. Then we show how to calculate a solution for $w(t)$ as a function of $z(t)$ in the special case in which $z(t)$ is the solution to a first-order differential equation system.

Consider first the matrix M_x. When $z(t)$ is set to zero for all t, the constraints are autonomous differential equations. Roberts (1971) and Denman and Beavers (1976) proposed a matrix sign algorithm for solving the resulting optimization problem and in particular for computing M_x. An attractive feature of the matrix sign algorithm is that it avoids computing the Jordan Decomposition of the Hamiltonian matrix H. Instead the matrix M_x is computed by calculating successively the average of a matrix and its inverse. The algorithm is initialized at the matrix H. More precisely, for any nonsingular matrix G, define $\mathcal{R}(G)$ via:

$$\mathcal{R}(G) = (1/2)\,[G + (G)^{-1}]\,. \tag{5.1}$$

Consider the sequence $\{\mathcal{R}^j(H)\}$ where $\mathcal{R}^j$ denotes the mapping $\mathcal{R}$ applied j times in succession. Using the Jordan decomposition for H, it follows that

$$\mathcal{R}^j(H) = E\mathcal{R}^j(J)E^{-1}\,. \tag{5.2}$$

Recall that J is a matrix with the eigenvalues of H on the diagonal. The first n diagonal entries contain the eigenvalues with strictly negative real parts and the remaining n diagonal entries contain the eigenvalues with strictly positive real parts. Some of the entries of J immediately to the right of eigenvalues that occur more than once are one and the remaining entries of J are zero. As a result, the sequence $\{\mathcal{R}^j(J)\}$ converges to a diagonal matrix with minus ones for the first n entries and ones for the second n entries.

Let H^∞ denote the limit point of $\{\mathcal{R}^j(H)\}$ and partition this matrix:

$$H^\infty = \begin{bmatrix} H_{11}^\infty & H_{12}^\infty \\ H_{21}^\infty & H_{22}^\infty \end{bmatrix}. \tag{5.3}$$

Recall that E is partitioned as

$$E = \begin{bmatrix} E_{11} & E_{12} \\ E_{21} & E_{22} \end{bmatrix}. \tag{5.4}$$

Similarly, partition $(E)^{-1}$ as

$$(E)^{-1} = \begin{bmatrix} E^{11} & E^{12} \\ E^{21} & E^{22} \end{bmatrix}. \tag{5.5}$$

Then

$$H_{11}^{\infty} = -E_{11}\,E^{11} + E_{12}\,E^{21} = -2E_{11}\,E^{11} + I\ , \tag{5.6}$$

and

$$H_{21}^{\infty} = -E_{21}\,E^{11} + E_{22}\,E^{21} = -2E_{21}\,E^{11}\ . \tag{5.7}$$

As long as E^{11} is nonsingular,

$$M_x = H_{21}^{\infty}\,(H_{11}^{\infty} - I)^{-1}\ . \tag{5.8}$$

Hence M_x can be approximated by computing $\mathcal{R}^j(H)$ for a sufficiently large value of j and then applying formula (5.8).

The restriction that E^{11} be nonsingular is not satisfied for some interesting parameter configurations of our model. For instance, E^{11} can be singular when it is not optimal for the state function x to be in L_+^n. When E^{11} is singular, we add a quadratic penalty $\hat{\delta}I$ to Ω_{xx} in the criterion for the *OLR* problem where $\hat{\delta}$ is strictly positive. This ensures that it is optimal for x to be in L_+^n. By selecting $\hat{\delta}$ to be sufficiently small, we can approximate the matrix M_x with arbitrary accuracy.

Anderson (1978) showed how to exploit the structure of the Hamiltonian matrix to simplify the iterations in the matrix sign algorithm. Let G be any Hamiltonian matrix. Anderson (1978) suggested the following partitioning:

$$\mathcal{R}(G)_{11} = (1/2)\ \{G_{11} + [G_{11} + G_{12}(G_{11}')^{-1}G_{21}]^{-1}\} \tag{5.9}$$

$$\mathcal{R}(G)_{12} = (1/2)\ \{G_{11} + [G_{11} + G_{12}(G_{11}')^{-1}G_{21}]^{-1}G_{12}\,(G_{11}')^{-1}\} \tag{5.10}$$

$$\mathcal{R}(G)_{21} = (1/2)\ \{G_{12} + (G_{11}')^{-1}G_{21}[G_{11} + G_{12}(G_{11}')^{-1}\,G_{21}]^{-1}\} \tag{5.11}$$

where $\mathcal{R}(G)_{11}$ is the upper right n by n block of $\mathcal{R}(G)$, and so on. This partitioning exploits the fact that the (2,2) block of the Hamiltonian matrix is the negative transpose of the (1,1) block. The matrix sign algorithm preserves this relation among the partitions of a matrix.

Consider next the computation of $w(t)$. For general specifications of $z(t)$, it is not possible to simplify further (4.33). We now impose some additional structure on $z(t)$. Suppose that $z(t)$ is the solution to the differential equation system:

$$Dz(t) = A_{zz}\,z(t) \tag{5.12}$$

where the eigenvalues of A_{zz} have real parts that are strictly negative. In this case, $z(t+\tau)$ is $\exp(A_{zz}\tau)z(t)$ implying that

$$w(t) = M_z\, z(t) \tag{5.13}$$

where

$$M_z = \int_0^\infty \exp[(H_{12}\, M_x + H_{11})'\tau]\, (K_2 - M_x\, K_1)\, \exp(A_{zz}\,\tau) d\tau\ . \tag{5.14}$$

There are two alternative ways of computing M_z. Using integration-by-parts, it can be shown that M_z satisfies:

$$(H_{12}\, M_x + H_{11})'\, M_z + M_z\, A_{zz} = K_2 - M_x\, K_1 \tag{5.15}$$

This equation is linear in the entries of the matrix M_z. Furthermore, since the eigenvalues of $(H_{12}\, M_x + H_{11})'$ and A_{zz} have strictly negative real parts, it can be shown that M_z is the unique solution to (5.15). Therefore, one way to compute M_z is to write (5.15) as a system of linear equations in the entries of M_z and solve that system of equations.

Following Denman and Beavers (1976), a second approach is to note that $w(t)$ and $z(t)$ satisfy the composite first-order differential equation system:

$$\begin{bmatrix} Dw(t) \\ Dz(t) \end{bmatrix} = \hat{H} \begin{bmatrix} w(t) \\ z(t) \end{bmatrix} \tag{5.16}$$

where

$$\hat{H} = \begin{bmatrix} -(H_{12}\, M_x + H_{11})' & (K_2 - M_x\, K_1) \\ 0 & A_{zz} \end{bmatrix}, \tag{5.17}$$

and to apply the matrix sign algorithm to find the stable solution to the composite system. Since $\hat{H}$ is upper triangular, the collection of eigenvalues of $\hat{H}$ are given by the union of the collection of eigenvalues of $-(H_{12}\, M_x + H_{11})'$ and the collection of eigenvalues of A_{zz}. The matrix M_z is found by mimicking the approach used to solve differential equation system (4.16). In particular, we deduce the limit point to the sequence $\{\mathcal{R}^j(\hat{H})\}$ and use the analog to formula (5.8) to compute M_z.

The matrix $\hat{H}$ is upper block triangular, and the matrix sign algorithm preserves this triangularity. These features can be exploited by partitioning the algorithm. Let $\hat{G}$ be an upper triangular matrix partitioned as

$$\hat{G} = \begin{bmatrix} \hat{G}_{11} & \hat{G}_{12} \\ 0 & \hat{G}_{22} \end{bmatrix}. \tag{5.18}$$

Then the matrix sign algorithm can be partitioned as:

$$\mathcal{R}(\hat{G})_{11} = (1/2)\ [\hat{G}_{11} + (\hat{G}_{11})^{-1}] \tag{5.19}$$

$$\mathcal{R}(\hat{G})_{12} = (1/2)\ [\hat{G}_{12} - (\hat{G}_{11})^{-1}\, \hat{G}_{12}\, (\hat{G}_{22})^{-1}] \tag{5.20}$$

$$\mathcal{R}(\hat{G})_{22} = (1/2)\ [\hat{G}_{22} + (\hat{G}_{22})^{-1}] \tag{5.21}$$

with $\mathcal{R}(\hat{G})_{21} = 0$. Finally, since the upper left block of $\hat{H}$ has eigenvalues with strictly positive real parts and the lower right block has eigenvalues with strictly negative real parts, it is straightforward to show that $\mathcal{R}^{\infty}(\hat{H})_{11} = I$ and that $\mathcal{R}^{\infty}(\hat{H})_{22} = -I$. The analog to (5.8) is simply

$$M_z = (-1/2)\mathcal{R}^{\infty}\,(\hat{H})_{12}\,. \tag{5.22}$$

When $z(t)$ satisfies (5.12), it is possible to augment the state vector $x(t)$ to include $z(t)$. An alternative solution approach is to form the Hamiltonian matrix for this augmented system and to initialize the matrix sign algorithm at this matrix. Although the approach suggested in this section requires two applications of the matrix sign algorithm, in each stage the algorithm is applied to matrices with smaller dimensions. As a consequence, the dual applications of the matrix sign algorithm typically will be faster than the single application to the augmented Hamiltonian matrix. This two-stage algorithm is the continuous time counterpart to a two-stage algorithm for solving discrete time quadratic control problems suggested by Hansen and Sargent (1981a).

6. Adjustment Cost Model of Investment

In this section we investigate the solution to the *OLR* problem in a special set of circumstances. We focus on competitive equilibrium models of investment under rational expectations as in Lucas (1981), Lucas and Prescott (1971) and Brock and MaGill (1979). The technologies we consider are the continuous time counterparts to technologies studied in Hansen and Sargent (1981a). There is a vector of capital stocks and there are costs to adjusting these stocks as in Lucas (1967), Gould (1968), Treadway (1969) and Mortenson (1973). The adjustment costs may be of higher order than one, which in the context of continuous time models may apply to higher derivatives of the capital stocks. The economic environment is described in Subsection 6.A. In Subsection 6.B we derive an alternative representation of the first-order conditions that can be interpreted as Euler equations. In Subsection 6.C we describe an alternative approach to solving the *OLR* problem and relate this approach to the one described in Section 5.

6a. Setup

Consider a special case of the general model presented in Section 1. Suppose there is no household capital so that (1.2) is replaced by

$$s(t) = \Pi c(t) \; . \tag{6.1}$$

Let $\hat{y}(t)$ be an m-dimensional vector of productive capital stocks at time t, and define $\hat{x}(t)' = k(t)' \equiv [D^{\ell-1}\hat{y}(t)', \; D^{\ell-2}\hat{y}(t)', \ldots, \hat{y}(t)']$ and $\hat{u}(t) \equiv D^{\ell}\hat{y}(t)$. In this case,

$$D\hat{x}(t) = \hat{A}\hat{x}(t) + B_u \hat{u}(t) \tag{6.2}$$

where

$$\hat{A} = \begin{bmatrix} 0 & 0 & \ldots & 0 & 0 \\ I & 0 & \ldots & 0 & 0 \\ \vdots & \vdots & \vdots & \vdots & \vdots \\ 0 & 0 & \ldots & I & 0 \end{bmatrix} \quad B_u = \begin{bmatrix} I \\ 0 \\ \vdots \\ 0 \end{bmatrix} . \tag{6.3}$$

Resource constraint (1.5) is imposed as before with $\hat{u}(t) = i(t)$. Notice that depreciation in capital is not reflected in specification of $\hat{A}$. This is because we have let the control be a measure of net investment. Suppose instead that $i(t)$ denotes a measure of gross investment and that

$$i(t) = \Gamma^* k(t) + \hat{u}(t) \tag{6.4}$$

where $\Gamma^* k(t)$ measures the investment required to replace any depreciated capital stock. Then (1.5) and (6.4) imply that

$$\Phi_c \, c(t) + \Phi_i \, \hat{u}(t) = (\Gamma - \Phi_i \Gamma^*)k(t) + f(t) \; . \tag{6.5}$$

Hence our use of net investment instead of gross investment as the control function requires that we replace Γ by $(\Gamma - \Phi_i \Gamma^*)$ in constraint (1.5).

6b. Euler Equations

Let $y(t) = \exp(-\epsilon t)\hat{y}(t)$. It is convenient to represent $x(t)$ in terms of $y(t)$ and its derivatives. Note that

$$Dy(t) = -\epsilon y(t) + \exp(-\epsilon t) D\hat{y}(t) \; . \tag{6.6}$$

In operator notation we have that

$$(D + \epsilon) y(t) = \exp(-\epsilon t) D\hat{y}(t) \; . \tag{6.7}$$

Similarly,

$$(D+\epsilon)^j y(t) = \exp(-\epsilon t) D^j \hat{y}(t) \tag{6.8}$$

where the operator raised to a power denotes sequential application of the operator. Hence we can represent $x(t)$ as:

$$x(t) = \begin{bmatrix} (D+\epsilon)^{\ell-1} I \\ (D+\epsilon)^{\ell-2} I \\ \vdots \\ I \end{bmatrix} y(t) . \tag{6.9}$$

We use this structure to obtain an alternative representation of the first-order conditions (4.6) and (4.7). Define a matrix function P of a complex variable:

$$P(\zeta) \equiv \begin{bmatrix} (\zeta+\epsilon)^{\ell} I \\ (\zeta+\epsilon)^{\ell-1} I \\ \vdots \\ I \end{bmatrix} , \tag{6.10}$$

First-order condition (4.6) can be expressed as

$$[\Omega_{uu}\, \Omega_{ux}]\, P(D) y(t) + \Omega_{uz}\, z(t) = -B_u'(-DI - A')^{-1} \lambda_s(t) . \tag{6.11}$$

Equations of system (4.7) can be expressed as

$$[\Omega_{xu}\, \Omega_{xx}]\, P(D) y(t) + \Omega_{xz}\, z(t) = \lambda_s(t) . \tag{6.12}$$

Recall that A' is given by

$$A' = \begin{bmatrix} -\epsilon I & I & 0 & \dots & 0 & 0 \\ 0 & -\epsilon I & I & \dots & 0 & 0 \\ \vdots & \vdots & \vdots & \vdots & \vdots & \vdots \\ 0 & 0 & 0 & \dots & 0 & -\epsilon I \end{bmatrix} \tag{6.13}$$

and that $B_u' = [I\ 0]$. The matrix operator $B_u'(-\zeta I - A')^{-1}$ consists of the first m rows of the operator $(-\zeta I - A')^{-1}$. Then

(6.14)
$$-B_u'(-\zeta I - A')^{-1} = -[(-\zeta+\epsilon)^{-1} I;\ (-\zeta+\epsilon)^{-2} I; \dots (-\zeta+\epsilon)^{-\ell} I] .$$

Substituting into (6.11) we obtain

$$
\begin{aligned}
&[\Omega_{uu}\,\Omega_{ux}]\,P(D)y(t) + \Omega_{uz}\,z(t) + \\
&\Big[(-D+\epsilon)^{-1} I\ ;\ (-D+\epsilon)^{-2} I\ ;\ \ldots\ (-D+\epsilon)^{-\ell} I\Big] \\
&\{\,[\Omega_{xu}\,\Omega_{xx}]\,P(D)y(t) + \Omega_{xz}\,z(t)\} = 0\ .
\end{aligned}
\tag{6.15}
$$

Define a matrix function Q of a complex variable ζ:

$$
\begin{aligned}
Q(\zeta) &\equiv [I\ ;\ (-\zeta+\epsilon)^{-1} I\ ;\ (-\zeta+\epsilon)^{-2} I\ ;\ \ldots\ (-\zeta+\epsilon)^{-\ell} I] \\
&= (-\zeta+\epsilon)^{-\ell}\, P(-\zeta)'\ .
\end{aligned}
\tag{6.16}
$$

Then (6.15) can be expressed as

$$
[Q(D)\Omega_{11} P(D)]y(t) = -Q(D)\Omega_{12}\, z(t)
\tag{6.17}
$$

where $\Omega_{11} \equiv \begin{bmatrix} \Omega_{uu} & \Omega_{ux} \\ \Omega_{xu} & \Omega_{xx} \end{bmatrix}$ and $\Omega_{12} = \begin{bmatrix} \Omega_{uz} \\ \Omega_{xz} \end{bmatrix}$.

In (6.17) $Q(D)$ is a *forward* convolution operator and $P(D)$ a *backwards* derivative operator.

Finally, it is of interest to disentangle the effect of scaling by $\exp(-\epsilon t)$. Hence we deduce a corresponding equation for $\hat{y}(t)$ in terms of $\hat{z}(t)$. It follows from (6.6) that

$$
Dy(t) = \exp(-\epsilon t)\ [(D-\epsilon)\hat{y}(t)]\ .
\tag{6.18}
$$

To undo the scaling in (6.17), we simply substitute $(D-\epsilon)$ for D and multiply by $\exp(\epsilon t)$:

$$
[Q(D-\epsilon)\Omega_{11}\, P(D-\epsilon)]\,\hat{y}(t) = -Q(D-\epsilon)\,\Omega_{12}\,\hat{z}(t)\ .
\tag{6.19}
$$

These are the Euler equations for the optimization problem.

6c. Solution

We now describe an alternative Euler equation approach to solving equations (6.19). In light of (6.16)

$$
Q(\zeta)\,\Omega_{11}\,P(\zeta) = P(-\zeta)'\,\Omega_{11}\,P(\zeta)/(-\zeta+\epsilon)^{\ell}\ .
\tag{6.20}
$$

Divide both sides of (6.20) by $(\zeta+\epsilon)^{\ell+1}\ (-\zeta+\epsilon)$ and evaluate the resulting function at $\zeta = i\theta$:

$$
\begin{aligned}
F(\theta) &\equiv P(-i\theta)'\,\Omega_{11} P(i\theta)/[(-i\theta+\epsilon)\,(i\theta+\epsilon)]^{\ell+1} \\
&= \Psi_{uu}(i\theta)/[(-i\theta+\epsilon)\,(i\theta+\epsilon)]
\end{aligned}
\tag{6.21}
$$

By virtue of Assumption 3.1, $F(\theta)$ is a positive definite matrix for all θ in R. In addition, F is a rational function of θ, $F(-\theta) = F(\theta)'$ and

$$\lim_{\theta \to \infty} F(\theta) = 0 . \tag{6.22}$$

As a consequence, F is the spectral density function for a linearly regular, stochastically nonsingular continuous time stochastic process. Since the spectral density function is rational, with common denominator $[(-i\theta + \epsilon)(i\theta + \epsilon)]^{\ell+1}$, it can be factored:

$$F(\theta) = \hat{P}(-i\theta)' V \hat{P}(i\theta) / [(-i\theta + \epsilon)(i\theta + \epsilon)]^{\ell+1} \tag{6.23}$$

where V is a nonsingular symmetric matrix, $\hat{P}(\zeta)$ is a polynomial of degree ℓ that is nonsingular in the left-half plane of C and

$$\lim_{\zeta \to \infty} \hat{P}(\zeta)/\zeta^{\ell} = I \tag{6.24}$$

(e.g. see Rozanov 1967). As a consequence, the composite operator $[Q(D)\Omega_{11}P(D)]$ used in Euler equation (6.16) can be factored

$$Q(D)\Omega_{11}P(D) = [\hat{P}(-D)/(-D + \epsilon)^{\ell}]' V \hat{P}(D) . \tag{6.25}$$

The operator $\hat{P}(D)$ is a backwards derivative operator and the operator $[\hat{P}(-D)/(-D + \epsilon)^{\ell}]$ is a forward convolution that can be expressed as the matrix linear combination of the identity operator and the forward convolution operators $1/(-D + \epsilon)$, $1/(-D + \epsilon)^2, \ldots 1/(-D + \epsilon)^{\ell}$.

The factorization given in (6.25) is of interest because the rational function $[\hat{P}(-\zeta)/(-\zeta + \epsilon)^{\ell}]$ is nonsingular in the left-half plane of C. As a consequence the forward operator $[\hat{P}(-D)/(-D + \epsilon)^{\ell}]$ has a one-sided forward inverse that can be characterized by inverting $[\hat{P}(-\zeta)/(-\zeta + \epsilon)^{\ell}]$. Let

$$\hat{Q}(\zeta) \equiv V^{-1}(-\zeta + \epsilon)^{\ell} \, \hat{P}(-\zeta)'^{-1} \, Q(\zeta)\Omega_{12} . \tag{6.26}$$

Then the entries of $\hat{Q}(\zeta)$ are rational functions of ζ with numerator orders that do not exceed the denominator orders. In addition, the poles of $\hat{Q}(\zeta)$ reside in the left-half plane of C. As a consequence $\hat{Q}(D)$ can be expressed as a matrix linear combination of the identity operator and convolution operators. The terms of this matrix linear combination can be deduced by computing a matrix partial fractions decomposition of $\hat{Q}(\zeta)$.

Exploiting these calculations, it follows that a recursive representation for the solution is given by

$$\hat{P}(D)y(t) = \hat{Q}(D)z(t) . \tag{6.27}$$

Removing the effect of discounting yields:

$$\hat{P}(D-\epsilon)\hat{y}(t) = \hat{Q}(D-\epsilon)\hat{z}(t) . \tag{6.28}$$

The operator $\hat{P}(D-\epsilon)$ determines the *feedback* part of the decision rule for $D^{\ell}y(t)$ as a function of $D^{j}y(t)$ for $j = 0, 1, \ldots, \ell - 1$, and the operator $\hat{Q}(D-\epsilon)$ determines the *feedforward* part of the decision rule involving $z(t)$ and forward convolution integrals of $z(t)$.

Recall from Section 5 that application of the matrix sign algorithm to the Hamiltonian matrix can be used to compute the coefficients of $\hat{P}(\zeta)$. Alternatively these coefficients can be calculated by factoring the spectral density F. The coefficients of the partial fractions decomposition for $\hat{Q}(\zeta)$ can be computed as in Hansen and Sargent (1981a) for the special case in which the zeros of $\det[\hat{P}(\zeta)]$ are distinct. Hansen and Sargent suggest an adaptation of an algorithm proposed by Emre and Huseyin (1975). When $z(t)$ is the solution to differential equation (5.12), the recursive methods described in Section 5 also deliver an expression for the convolution integral $\hat{Q}(D)z(t)$ as a linear function of $z(t)$.

7. Permanent Income Model

In this section we consider continuous-time versions of the permanent income model of consumption as studied by Hall (1978), Flavin (1981), and Sargent (1987b). All of these authors assumed that preferences are separable over time. As in Hansen (1987) and Heaton (1989), we extend the permanent income model to accommodate preferences that are not time separable.

The technology for this economy is particularly simple. There is a single consumption good and a single physical capital stock. For simplicity we assume that the capital stock does not depreciate ($\Delta_k = 0$ and $\Theta_k = 1$). At each point in time output is obtained from capital $k(t)$ and an endowment $f(t)$. The instantaneous return Γ on capital is set equal to the subjective rate of time preference ρ. Output is divided between consumption and investment with $\Phi_c = \Phi_i = 1$.

We consider a variety of alternative specifications of the household technology. In all cases there is a single household capital stock and a single consumption service. Household capital depreciates over time so that $\Delta_h < 0$. In order that the household capital can be interpreted as a weighted average of current and past consumption, we let $\Theta_h = -\Delta_h$. Current consumption and the household capital stock are used to produce services via relation (1.2). The scalar $b(t)$ determines the satiation point in consumption services at time t. In our discussion of this model we will assume that $b(t)$ is sufficiently big that the marginal utility of services is positive.

Consider first the case in which $\Lambda = 1$ and $\Pi = 0$. In this case the consumption good is durable, and $h(t)$ measures the stock of durable goods. The services in each time period are simply equal to the stock of durable goods. Unfortunately, under this specification, the model does not satisfy one of the maintained assumptions of our analysis. In particular, $\Omega_{uu} = 0$ and hence Assumption 4.1 is violated. Following Heaton (1989) we redefine the problem in terms of accumulated quantities. For example, accumulated consumption at time t, $c^a(t)$ is given by:

$$c^a(t) = \int_0^t c(\tau) d\tau \; . \tag{7.1}$$

The resource constraint (1.5) is then interpreted as holding over accumulated quantities. Ignoring initial conditions, the quantities in the laws of motion (1.3) and (1.4) are also replaced by their accumulated counterparts. The mapping between the stock of the durable good and services given by (1.2) must be replaced, however, in order to maintain the original preference ordering over consumption. Services at time t are now given by:

$$s(t) = \Delta_h \, h^a(t) - \Delta_h \, c^a(t) + \exp(\Delta_h t) \mu_h \tag{7.2}$$

Hence in the transformed problem Λ is replaced by Δ_h and $\Pi = -\Delta_h$. The term $\exp(\Delta_h t)\mu_h$ can be absorbed into the forcing function $b(t)$. The results in this paper can then be immediately applied to the accumulated functions.

To interpret the solution in terms of the original functions, time derivatives must be taken of the solutions for the accumulated functions. For this example, however, ordinary time derivatives do not exist for all of the functions. In these cases the resulting time derivatives must be viewed as generalized derivatives. In Heaton (1989), these issues are discussed in more detail.

When Π is strictly positive, the original formulation of the model still applies. In this case current consumption and the household capital stock are substitutes in the production of consumption services. Although the consumption good is durable, additional services are generated by new acquisitions of the goods.

Habit persistence in preferences over consumption as examined for example by Pollak (1970), Ryder and Heal (1973), Sundaresan (1989), Constantinides (1990) and Heaton (1989) can be accommodated by assuming that $\Lambda = -1$, $\Pi \geq 1$. In this case consumption and capital are complements in the production of consumption services.[1]

When $\Lambda = 0$ and $\Pi = 1$, the induced preferences for consumptions are time separable. The solution method described in Section 6 can be applied to characterize the optimal law of motion for capital. In this case, we can drop the household capital stock, and the productive capital stock becomes the only component of the endogenous state variable. The restriction that $k(t)$ discounted by $\exp(-\epsilon t)$ be in L^1_+ is important in obtaining a solution that is of interest. Without this restriction, $c(t)$ is set to $b(t)$ which in turn is supported by an unstable time path for the discounted capital stock.

To solve this model we let $z(t)' = [b(t), f(t)]$. The matrices Ω_{11} and Ω_{12} of Section 6 are given by:

$$\Omega_{11} = \begin{bmatrix} 1 & -\rho \\ -\rho & \rho^2 \end{bmatrix} \quad \text{and} \quad \Omega_{12} = \begin{bmatrix} -1 \\ \rho \end{bmatrix} [-1, 1] \,. \tag{7.3}$$

The composite operator $Q(D)\Omega_{11}P(D)$ in this case is given by:

$$[1 \; 1/(-D+\epsilon)] \begin{bmatrix} 1 & -\rho \\ -\rho & \rho^2 \end{bmatrix} \begin{bmatrix} (D+\epsilon) \\ 1 \end{bmatrix} = (-D+\epsilon)\,(D+\epsilon)/(-D+\epsilon) \,. \tag{7.4}$$

Hence the operator $\hat{P}(D)$ of equation (6.23) is $(D+\epsilon)$ and V is equal to 1. The operator $-Q(D)\Omega_{12}$ is given by:

$$\begin{aligned} -Q(D)\Omega_{12} &= -[1 \; 1/(-D+\epsilon)] \begin{bmatrix} -1 \\ \rho \end{bmatrix} [-1, 1] \\ &= [1 - \rho/(-D+\epsilon)]\,[-1, 1] \,. \end{aligned} \tag{7.5}$$

The operator $\hat{Q}(D)$ of equation (6.26) is

$$\begin{aligned} \hat{Q}(D) &= (-D+\epsilon)\,[1 - \rho/(-D+\epsilon)]\,[-1, 1]/(-D+\epsilon) \\ &= [1 - \rho/(-D+\epsilon)]\,[-1, 1] \,. \end{aligned} \tag{7.6}$$

It follows from (6.28) that the capital stock satisfies:

(7.7)
$$\begin{aligned} Dk(t) &= [1 - \rho/(-D + \epsilon)]\,[-1,\, 1]\, z(t) \\ &= [1 - \rho/(-D + \epsilon)]\,[f(t) - b(t)] \\ &= [f(t) - b(t)] - \rho \int_0^\infty \exp(-\rho\tau) - [f(t+\tau) - b(t+\tau)]d\tau \,. \end{aligned}$$

In other words, investment is calculated by comparing the current level of the endowment relative to the satiation point to a weighted average of future endowments relative to satiation points. Heaton (1989) displays solutions to the model for other settings for Λ and Π.

Appendix A

In this appendix we verify that formulas (2.3), (2.6), (2.8), and (2.15) apply to all of L_2^n. Let 1_τ denote the indicator function of the set $[-\tau,\, \tau]$, and let x denote any member of L_2^n. Then $1_\tau x$ is in $L_1^n \cap L_2^n$ for each t and $\{1_\tau x : t \geq 1\}$ converges in L_2^n to x. Since T is continuous, $\{T(1_\tau x) : t \geq 1\}$ converges in L_2^n to $T(x)$. Let O be any of the four operators introduced in Section 2, and let ϕ be the corresponding function such that $\phi T(x) = T[O(x)]$. The function ϕ is bounded in all four cases. Therefore, $\{T[O(1_\tau x)] : \tau \geq 1\}$ converges in L_2^n to $T[O(x)]$. The Parseval formula implies that $\{O(1_\tau x) : \tau \geq 1\}$ converges in L_2^n to $O(x)$. A subsequence $\{O[1_{\tau(j)} x] : j \geq 1\}$ converges pointwise to $O(x)$ except on a set of measure zero. Formulas (2.3), (2.6), (2.8) and (2.15) involve integral representations. In this case, $O(1_\tau x)$ has an integral representation. Since x can be expressed as a linear combination of nonnegative, vectors of real-valued functions, it follows from the Monotone Convergence Theorem that $O(x)$ has the same integral representation.

Appendix B

In this appendix we establish that there exists a solution to the *OLR* problem when Assumption 4.1 is satisfied. Since the matrix Ω is positive semidefinite, the criterion function is always less than or equal to zero. Define

$$(B.1) \qquad T(u,\, x) \equiv \int_0^\infty [u(t)'\; x(t)'\; z(t)']\Omega \begin{bmatrix} u(t) \\ x(t) \\ z(t) \end{bmatrix} dt \,,$$

and put

$$(B.2) \quad \bar{\delta} \equiv \inf\{T(u,\, x) : u \in L_+^m,\; x \in L_+^n,\; (u,\, x) \text{ satisfies } (1.13)\} \,.$$

Then $\bar{\delta}$ is finite and there exists a sequence $\{(u_j, x_j)\}$ such that $\{T(u_j, x_j)\}$ converges to $\bar{\delta}$, $u_j \in L_+^m$, $x_j \in L_+^n$, and (u_j, x_j) satisfies (1.13). For any positive integers j and ℓ,
(B.3)

$$T(u_j - u_\ell, x_j - x_\ell) + T(u_j + u_\ell, x_j + x_\ell) = 2T(u_j, x_j) + 2T(u_\ell, x_\ell)\,.$$

Since (u_j, x_j) and (u_ℓ, x_ℓ) satisfy (1.13), $[(u_j + u_\ell)/2,\ (x_j + x_\ell)/2]$ satisfies (1.13). Consequently,

$$(B.4)\quad (1/4)T(u_j - u_\ell, x_j - x_\ell) \le (1/2)T(u_j, x_j) + (1/2)T(u_\ell, x_\ell) - \bar{\delta}\,.$$

Multiplying by four and taking limits, we obtain

$$(B.5)\qquad \lim_{j,\ell\to\infty} T(u_j - u_\ell, x_j - x_\ell) = 0\,.$$

The Parseval formula implies that

$$(B.6)\qquad \begin{aligned} T(u_j - u_\ell, x_j - x_\ell) &= (1/2\pi)\int_R [T(u_j - u_\ell)(\theta)]^{*\prime}\, \Psi_{uu}(i\theta) \\ &\qquad T(u_j - u_\ell)(\theta)d\theta \\ &\ge \delta(1/2\pi)\int_R [T(u_j - u_\ell)(\theta)\,|^2\, d\theta \\ &= \delta\int_0^\infty |\,(u_j(t) - u_\ell(t)\,|^2\, dt \end{aligned}$$

since Assumption 4.1 is satisfied. Consequently, $\{u_j\}$ is Cauchy in L_+^m. Since C is a continuous function mapping L_+^n into L_+^n, $\{x_j\}$ is Cauchy in L_+^n. Let u_s and x_s be the limit points of these two Cauchy sequences. Then (u_s, x_s) satisfies (3.12) and $T(u_s, x_s) = \delta$. Therefore, (u_s, x_s) solves the *OLR* Problem.

Appendix C

In this appendix we show that the matrix M_x is positive semidefinite and show that it can be used to represent the value function. We consider a finite horizon optimization problem. For convenience, we set the interval of time equal to one. In this case we modify the space L_+^1 of real-valued Borel measurable functions defined on [0, 1]. Instead of using Lebesgue measure on [0, 1] we augment the measure by adding an atom at the point {1} with unit measure. We then consider the following optimization problem

$$(C.1)\quad \max_{u\in L_+^n i,\, x\in L_+^n} (-1/2)\Big\{\int_0^1 [u(t)'\ x(t)']\Omega_{11}\begin{bmatrix} u(t)\\ x(t)\end{bmatrix} dt + x(1)'\mathcal{B}x(1)\Big\}$$

subject to $x(t) = \exp(At)\mu + \int_0^t \exp(A\tau)B_u u(t-\tau)d\tau$ for $0 \le t \le 1$. In (C.1) $\mathcal{B}$ is a positive semidefinite matrix. The first-order conditions for this optimization problem are essentially the same as for the *OLR* problem of Section 4 except that time period one must receive special attention. When we differentiate with respect to $x(1)$ we get that

$$\mathcal{B}x_s(1) = \lambda_s(1) . \tag{C.2}$$

The first-order conditions for $u(t)$ and $x(t)$ for $0 \le t \le 1$ are now given by

$$\begin{aligned}&\Omega_{uu}u_s(t) + \Omega_{ux}x_s(t)\\ &\qquad + B'\int_t^1 \exp[A'(\tau - t)]\lambda_s(\tau)d\tau + B'\exp[A'(1-t)]\lambda_s(1) = 0\end{aligned} \tag{C.3}$$

$$\Omega_{xu}\, u_s(t) + \Omega_{xx}\, x_s(t) = \lambda_s(t) \tag{C.4}$$

In this case we define the co-state function as

$$x_c(t) \equiv \int_t^1 \exp[A'(\tau - t)]\lambda_s(\tau)d\tau + \exp[A'(1-t)]\lambda_s(1) . \tag{C.5}$$

Notice that $x_c(1) = \lambda_s(1)$, and hence (C.2) and (C.5) imply that

$$\mathcal{B}x_s(1) = x_c(1) . \tag{C.6}$$

Consequently, we solve this finite horizon problem by solving the same differential equation system except that in addition to the initial condition that $x_s(0) = \mu$ we have the terminal condition (C.6).

Suppose that $\mathcal{B}$ is defined to be the positive semidefinite matrix such that the time t value function for the time-invariant infinite horizon problem is given by $-(1/2)x(t)'\mathcal{B}x(t)$. In this case the solution to the finite horizon and infinite horizon problems must agree as must the optimal state and co-state functions. As noted in the text, the solution to the infinite horizon problem satisfies:

$$M_x\, x_s(t) = x_c(t) \quad \text{for} \quad t \ge 0 . \tag{C.7}$$

By altering the time horizon of the finite-time problem, we have that

$$\mathcal{B}\, x_s(t) = x_c(t) \quad \text{for} \quad t \ge 0 . \tag{C.8}$$

Since (C.7) and (C.8) must hold for any arbitrary initial condition μ for $x(0)$, it follows that $\mathcal{B} = M_x$. Therefore, M_x is positive semidefinite and can be used to construct the value function for an optimization problem in which $z(t) = 0$ for $t \ge 0$.

Notes

1. For some values of $\Pi < 1$, the household technology generates *rational addiction* as suggested by Stigler and Becker (1977) and Becker and Murphy (1988).

8

Prediction Formulas for Continuous Time Linear Rational Expectations Models

by Lars Peter HANSEN and Thomas J. SARGENT

In this note we derive optimal prediction formulas to be used in solving continuous time rational expectations models. In these derivations we employ Laplace transforms in a manner analogous to the use of z transforms for solving discrete time optimal prediction problems in Hansen and Sargent (1980a, Appendix A). The formulas are intended to play the same role for continuous time models that the discrete time formulas for optimal predictions of geometric distributed leads did in Hansen and Sargent (1980a).

1. Convolutions and Prediction

Let L^1 and L^2 denote the spaces of all real-valued Borel measurable functions ϕ on $\mathbf{R}$ that are absolutely integrable and square integrable, respectively. Let W denote a random measure defined on $\mathbf{R}$ with increments that are orthogonal and second-moment stationary. In other words,

$$E\Big[W\{[t_2,\ t_1)\}^2\Big] = t_2 - t_1 \ \text{ for } \ t_2 > t_1 \ , \tag{1.1}$$

and

$$E\Big[W\{[t_4,\ t_3)\}W\{[t_2,\ t_1)\}\Big] = 0 \ \text{ for } \ t_4 > t_3 > t_2 > t_1 \ . \tag{1.2}$$

Using functions in L^2 and the random measure W, we construct second-moment stationary processes as convolutions:

$$x(t) = \int_{-\infty}^{+\infty} \phi(\tau)dW(t-\tau) \ . \tag{1.3}$$

The stochastic integral in (1.3) can be interpreted as the limit point of a mean-square convergent sequence of random variables (e.g. see Rozanov 1967). Relation (1.3) gives a convenient mapping between the space L^2 of functions and the space X of stochastic processes. It turns out that inner products on these two spaces coincide. More precisely, let ϕ_1 and ϕ_2 be any two functions in L^2. An implication of (1.1) and (1.2), is

$$\int_{-\infty}^{+\infty} \phi_1(\tau)\phi_2(\tau)d\tau = E[x_1(t)\, x_2(t)] \tag{1.4}$$

where x_1 and x_2 are given by convolution (1.3) using ϕ_1 and ϕ_2 respectively.

Let L_+^2 denote the subspace of L^2 consisting of all functions that are zero on $(-\infty, 0)$ and let L_-^2 denote the subspace of all functions that are zero on $[0, \infty)$. Clearly, L_+^2 and L_-^2 are orthogonal and $L^2 = L_-^2 \oplus L_+^2$. Any ϕ in L^2 can be decomposed uniquely into the sum of two functions $\phi^+ \in L_+^2$ and $\phi^- \in L_-^2$ via:[1]

$$\begin{aligned} \phi^+(t) &\equiv \begin{cases} \phi(t) & t \geq 0 \\ 0 & t < 0 \end{cases} \\ \phi^-(t) &\equiv \begin{cases} 0 & t \geq 0 \\ \phi(t) & t < 0\,. \end{cases} \end{aligned} \tag{1.5}$$

To formulate the prediction problems of interest, we use the random measure W to induce a family of information sets indexed by calendar time. Let $H(t)$ denote the space of random variables $x(t)$ given by (1.3) for ϕ's restricted to be in L_+^2. It follows from (1.4) that since L_+^2 is a Hilbert space, so is $H(t)$. Furthermore, the family of Hilbert spaces $\{H(t)\}$ is increasing in the sense that if $t_2 > t_1$, then $H(t_2) \supset H(t_1)$. Since $H(t)$ is constructed using the random measure W, the least squares projection operator $P[\cdot|H(t)]$ onto the space $H(t)$ is given by

$$P\Big[\int_{-\infty}^{+\infty} \phi(\tau)dW(t-\tau) \mid H(t)\Big] = \int_{-\infty}^{+\infty} \phi^+(\tau)dW(t-\tau)\,. \tag{1.6}$$

Hence the prediction process obtained by taking a process $x \in X$ constructed as a convolution of ϕ and dW and projecting it onto $H(t)$ for each t is a convolution of ϕ^+ and dW for ϕ^+ given in (1.5).

2. Transforms

One convenient way to represent functions in L^2 and characterize mapping (1.5) involves the use of transforms. For instance, Fourier transforms are valuable in characterizing the second moment properties of processes in X. For any ϕ in $L^1 \cap L^2$, the Fourier transform of ϕ is defined to be

$$\mathcal{F}t\,(\phi)(\theta) \equiv \int_{-\infty}^{+\infty} \exp(-i\theta t)\phi(t)dt\,. \tag{2.1}$$

There is a well known extension of $\mathcal{F}t$ from $L^1 \cap L^2$ to L^2. Using this extension, the spectral density function for x generated via (1.3) is just $|\mathcal{F}t\,(\phi)|^2$.

To characterize the implied second moment properties of the solutions to prediction problems of the form (1.6), we use Laplace transforms. These transforms are defined as follows. For any ϕ in L^2 and any ρ in $\mathbf{R}$ we construct a new function $\exp(-\rho t)\phi$. This new function may or may not be in L^2 depending on the value of ρ. Whenever it is in L^2, we define the Laplace transform to be:

$$\mathcal{L}p\,(\phi)(\mathbf{c}) \equiv \mathcal{F}t\,[\exp(-\rho t)\phi](\theta) \tag{2.2}$$

where $\mathbf{c} \equiv \rho + i\theta$.

The question of interest is the following. Given the Laplace transform $\mathcal{L}p\,(\phi)$ of a function $\phi \in L^2$, how can we compute or characterize $\mathcal{L}p\,(\phi^+)$ where ϕ^+ is defined in (1.5)? To answer this question, we first study Laplace transforms of functions $\phi \in L^2_+$. For any such ϕ, $\exp(-\rho t)\phi$ is also in L^2_+ as long as $\rho > 0$. Hence the Laplace transform $\mathcal{L}p\,(\phi)(\mathbf{c})$ is well defined on the closed right plane $\mathbf{C}_0^+$ where $\mathbf{C}_\delta^+ \equiv \{\mathbf{c} \in \mathbf{C} : \text{real}(\mathbf{c}) \geq \delta\}]$. Moreover, $\mathcal{L}p\,(\phi)$ is analytic in the interior of $\mathbf{C}_0^+$ (relative to $\mathbf{C}$). For $\delta > 0$ and $\mathbf{c} \in \mathbf{C}_\delta^+$,

$$\begin{aligned} |\mathcal{L}p\,(\phi)(\mathbf{c})| &\leq \int_0^\infty |\phi(t)| \exp(-\delta t)dt \\ &\leq \left[\int_0^\infty |\phi(t)|^2 dt \int_0^\infty \exp(-2\delta t)dt\right]^{1/2} \\ &\leq \left[\int_0^\infty |\phi(t)|^2 dt/2\delta\right]^{1/2} \end{aligned} \tag{2.3}$$

where the second inequality is an application of the familiar Cauchy-Schwarz Inequality. The right side of (2.3) gives a uniform bound (in

c) on $\mathcal{L}p(\phi)$ over the set $\mathbf{C}_\delta^+$. This bound becomes arbitrarily small as δ tends to plus infinity.

Consider next functions $\phi \in L_-^2$. For any such ϕ, $\mathcal{L}p(\phi)(\mathbf{c})$ is always well defined for $\mathbf{c}$ in left half plane $\mathbf{C}_0^- \equiv \{\mathbf{c} \in \mathbf{C} : \text{real}(\mathbf{c}) \leq 0\}$, and $\mathcal{L}p(\phi)$ is analytic in the interior of that domain. Define $\mathbf{C}_\delta^- \equiv \{\mathbf{c} \in \mathbf{C} : \text{real}(\mathbf{c}) \leq \delta\}$. Mimicking the previous argument, it can be shown that for any $\delta < 0$, $\mathcal{L}p(\phi)$ is bounded on the domain $\mathbf{C}_\delta^-$ and that the bound can be made arbitrarily small by driving δ towards minus infinity.

For general functions ϕ in L^2, $\mathcal{L}p(\phi)$ may only be defined on the imaginary axis, i.e. for real$(\mathbf{c}) = 0$. We are interested in a smaller class of functions, however. Let Φ be the set of all functions $\phi \in L^2$ such that $\mathcal{L}p(\phi^-)$ is analytic in the interior of a region $\mathbf{C}_\delta^-$ for some $\delta > 0$. In this case $\mathcal{L}p(\phi)$ is analytic in the interior of the strip $\mathbf{C}_\delta^- \cap \mathbf{C}_0^+$. Furthermore, for any closed interval $J \subset (0, \delta)$, $\mathcal{L}p(\phi)$ is bounded on $\{\mathbf{c} \in \mathbf{C} : \text{real}(\mathbf{c}) \in J\}$. Define $\mathcal{A}$ to be the collection of all Laplace transforms of functions $\phi \in \Phi$.

The following result gives the decomposition for $a \in \mathcal{A}$ corresponding to the decomposition $\phi = \phi^+ + \phi^-$.

Lemma: For any $a \in \mathcal{A}$ there is a unique decomposition $a = a^+ + a^-$ where

(i) a^- is analytic in the interior of $\mathbf{C}_\delta^-$, uniformly bounded on any closed half plane $\mathbf{C}_\rho^-$ for $\rho < \delta$ and

$$\lim_{\rho \to -\infty} \max_{\mathbf{c} \in \mathbf{C}_\rho^-} |a^-(\mathbf{c})| = 0 \; ;$$

(ii) a^+ is analytic in the interior of $\mathbf{C}_0^+$, uniformly bounded on any closed half plane $\mathbf{C}_\rho^+$ for any $\rho > 0$.

Proof: Functions a^- and a^+ satisfying (i) and (ii) are obtained by letting $a^- = \mathcal{L}p(\phi^-)$ and $a^+ = \mathcal{L}p(\phi^+)$. To show that the decomposition is unique, we let $a = b^+ + b^-$ be any other decomposition where b^- satisfies (i) and b^+ satisfies (ii). Note that

$$a^+ - b^+ = b^- - a^-$$

at least in the interior of the strip $\mathbf{C}_\delta^- \cap \mathbf{C}_0^+$. Since $a^+ - b^+$ is analytic in the interior of $\mathbf{C}_0^+$ and $b^- - a^-$ is analytic in the interior of $\mathbf{C}_\delta^-$, $b^- - a^-$ can be extended to be analytic on all of $\mathbf{C}$. Furthermore, the uniform bounds on $a^+ - b^+$ and $b^- - a^-$ on overlapping half planes ensure that the extension of $b^- - a^-$ is bounded as well. The only functions that are bounded and analytic on **C** are constant. Since a^- and b^- satisfy (i) and the extension of $b^- - a^-$ to C is constant, $a^+ - b^+$ must be identically zero. ▮

Decompositions like that given in the Lemma apply to a much more general collection of analytic functions than the Laplace transforms of functions in L^2. For instance, they also apply to Laplace transforms of generalized functions (e.g. see Beltrami and Wohlers 1966). However, these more general decompositions may not be unique. For instance, suppose we ignore the requirement

$$\lim_{\rho \to -\infty} \max_{\mathbf{c} \in \mathbf{C}_\rho^-} |a^-(\mathbf{c})| = 0$$

in (i) of the Lemma. Then one can always add complex numbers to a^+ and subtract the same numbers from a to obtain other decompositions of a. If in addition, we ignore the bound restrictions in (i) and (ii) of the Lemma then one can add functions, such as polynomials, that are analytic in the entire complex plane to a^+ and subtract them from a^- to obtain other decompositions of a. Therefore in applying the Lemma to compute $\mathcal{L}p(\phi^+)$, it is important to check whether the candidates for $\mathcal{L}p(\phi^+)$ and $\mathcal{L}p(\phi^-)$ satisfy the bounds restrictions in (i) and (ii).

3. Examples

We now apply the Lemma to obtain frequency domain characterizations of the solutions to prediction problems that occur in rational expectations models. These problems all have the following structure. Let $\psi \in L_+^2$, and define $y(t)$ by the convolution:

$$y(t) = \int_0^{+\infty} \psi(\tau) dW(t-\tau) .$$

Construct a new process by forming a *forward-looking* convolution using a function $\gamma \in L_-^1$: [2]

$$x(t) = \int_{-\infty}^{+\infty} \gamma(\tau) y(t-\tau) d\tau \equiv \int_{-\infty}^{+\infty} \phi(\tau) dW(t-\tau) \tag{3.1}$$

where ϕ is given by the convolution:

$$\phi(\tau) = \int_{-\infty}^{+\infty} \gamma(s) \psi(\tau - s) ds .$$

Applying the well known product representation for Fourier transforms of convolutions, we have that

$$\mathcal{F}t(\phi) = \mathcal{F}t(\psi) \mathcal{F}t(\gamma) .$$

This same result extends to Laplace transforms on the common domain of $\mathcal{L}p(\phi)$ and $\mathcal{L}p(\gamma)$. For the examples we consider, there will exist a $\delta > 0$ such that $\mathcal{L}p(\gamma)$ is defined on the interior of $\mathbf{C}_\delta^-$. Hence on the interior of the strip $\mathbf{C}_\delta^- \cap \mathbf{C}_0^+$,

$$\mathcal{L}p(\phi) = \mathcal{L}p(\psi)\mathcal{L}p(\gamma) . \tag{3.2}$$

We now investigate three related examples.

Example 1:

Suppose that γ is given by

$$\gamma(t) = \begin{cases} 0 & t \geq 0 \\ \exp(\delta t) & t < 0 \end{cases} . \tag{3.3}$$

Then

$$\mathcal{L}p(\gamma)(\mathbf{c}) = \int_{-\infty}^{0} \exp[(\delta - \mathbf{c})t]dt = 1/(\mathbf{c} - \delta)$$

for real$(\mathbf{c}) < \delta$. Thus

$$\mathcal{L}p(\phi)(\mathbf{c}) = \mathcal{L}p(\psi)(\mathbf{c})/(\mathbf{c} - \delta) .$$

Note that $\mathcal{L}p(\phi)$ is analytic on $\mathbf{C}$ except possibly at the point δ where it may have a pole. If $\mathcal{L}p(\psi)(\delta)$ is zero, the singularity at δ is removable and $\mathcal{L}p(\phi^+) = \mathcal{L}p(\phi)$. Usually $\mathcal{L}p(\phi)$ will have a pole at δ, and to compute $\mathcal{L}p(\phi^+)$ we must eliminate this pole. One candidate for $\mathcal{L}p(\phi^+)$ is

$$a^+(\mathbf{c}) = [\mathcal{L}p(\psi)(\mathbf{c}) - \mathcal{L}p(\psi)(\delta)]/(\mathbf{c} - \delta) .$$

Notice that the singularity of a^+ at δ is removable. The corresponding choice of a^- is

$$a^-(\mathbf{c}) = a(\mathbf{c}) - a^+(\mathbf{c}) = \mathcal{L}p(\psi)\,(\delta)/(\mathbf{c} - \delta) .$$

It is straightforward to show that a^+ and a^- satisfy the requirements of the Lemma. Therefore,

$$\mathcal{L}p(\phi^+) = [\mathcal{L}p(\psi)(\mathbf{c}) - \mathcal{L}p(\psi)(\delta)]/(\mathbf{c} - \delta) . \tag{3.4}$$

Formula (3.4) is the continuous time counterpart to formula (5) in Hansen and Sargent (1980a).

Example 2

More generally, suppose

$$\mathcal{L}p(\gamma)(\mathbf{c}) = p_n(\mathbf{c})/p_d(\mathbf{c})$$

where p_n and p_d are finite-order polynomials with real coefficients. To ensure that $p_n(\mathbf{c})/p_d(\mathbf{c})$ is the Laplace transform of a function in L^2_-, we assume that the order of p_d exceeds the order of p_n and that the zeros of p_d are in the interior of $\mathbf{C}_0^+$. In this case

$$\mathcal{L}p(\phi)(\mathbf{c}) = \mathcal{L}p(\psi)(\mathbf{c})p_n(\mathbf{c})/p_d(\mathbf{c}) ,$$

which has poles in the interior of $\mathbf{C}_0^+$ only at the zeros of p_d. Let a_j denote the principal part of the Laurent series expansion of $\mathcal{L}p(\phi)(\mathbf{c})$ at the jth zero of p_d. It follows from the partial fractions decomposition of a meromorphic function that

$$a^+ = \mathcal{L}p(\phi) - \sum_j a_j$$

is analytic in the interior of $\mathbf{C}_0^+$. Furthermore, the principal parts, a_j, are each sums of reciprocals of first and higher-order polynomials and hence satisfy

$$\lim_{\rho \to -\infty} \max_{\mathbf{c} \in \mathbf{C}_\rho^-} |a_j(\mathbf{c})| = 0 .$$

for each j. By construction,

$$a^- = \sum_j a_j$$

satisfies (i) of the Lemma where δ is the real part of the zero of p_n closest to the imaginary axis and is bounded on $\mathbf{C}_\rho^-$ for any $\rho < \delta$. Therefore, we have the following generalization of (3.4):

$$\mathcal{L}p(\phi^+) = \mathcal{L}p(\phi) - \sum_j a_j .$$

Example 3

Suppose that γ is given by (3.2), and $\mathcal{L}p(\psi)$ is a rational function:

$$\mathcal{L}p(\psi) = q_n/q_d$$

where q_n and q_d are polynomials with real coefficients. To guarantee that q_n/q_d is the Laplace transform of a function in L^2_+, we assume that the order of q_d exceeds the order of q_n and that the zeros of q_d are in the interior of $\mathbf{C}_0^-$. Solution (3.4) now becomes

$$\mathcal{L}p(\phi)(\mathbf{c}) = q_n(\mathbf{c})/[q_d(\mathbf{c})(\mathbf{c} - \delta)] .$$

From Example 1, we know that

$$\begin{aligned} \mathcal{L}p(\phi^+)(\mathbf{c}) &= [q_n(\mathbf{c})/q_d(\mathbf{c}) - q_n(\delta)/q_d(\delta)]/(\mathbf{c} - \delta) \\ &= [q_n(\mathbf{c})q_d(\delta) - q_n(\delta)q_d(\mathbf{c})]/[q_d(\mathbf{c})q_d(\delta)(\mathbf{c} - \delta)] . \end{aligned} \tag{3.5}$$

The right side of (3.5) has a removable singularity at δ by construction. This is evident because the polynomial $[q_n(\mathbf{c})q_d(\delta) - q_n(\delta)q_d(\mathbf{c})]$ has a zero at δ. Canceling the common factor $(\mathbf{c} - \delta)$ in the numerator and denominator results in

$$\mathcal{L}p(\phi^+)(\mathbf{c}) = q_n^+(\mathbf{c})/q_d^+(\mathbf{c})$$

where

$$q_d^+(\mathbf{c}) = q_d(\mathbf{c})q_d(\delta)$$

and q_n^+ satisfies

$$q_n^+(\mathbf{c})(\mathbf{c} - \delta) = [q_n(\mathbf{c})q_d(\delta) - q_n(\delta)q_d(\mathbf{c})] . \tag{3.6}$$

By equating coefficients of the polynomials on both sides of (3.6), one can construct a linear system of equations in the coefficients of $q_n^+(\mathbf{c})$. In fact there is a recursive structure to this equation system that can be exploited as follows. Let η_j denote the coefficient on $\mathbf{c}^j$ in $[q_n(\mathbf{c})q_d(\delta) - q_n(\delta)q_d(\mathbf{c})]$ and let ϵ_j denote the corresponding coefficient in $q_n^+(\mathbf{c})$. Then

$$-\delta\epsilon_0 = \eta_0$$

and

$$\epsilon_{j-1} - \delta\epsilon_j = \eta_j \quad \text{for } j \geq 1$$

which can be solved recursively beginning with ϵ_0. The solution to this recursion gives a continuous time counterpart to formulas reported in Hansen and Sargent (1980a, 1981b) for autoregressive and autoregressive moving-average processes.

4. Vector Information Structures

Suppose that W is an k-dimensional vector random measure with second moment stationary increments. We now replace (1.1) and (1.2) with

$$E\big[W\{[t_2, t_1)\}W\{[t_2, t_1)\}'\big] = (t_2 - t_1)I_k \quad \text{for } t_2 > t_1 ,$$

and

$$E\left[W\{[t_4, t_3)\}W\{[t_2, t_1)\}'\right] = 0 \quad \text{for } t_4 > t_3 > t_2 > t_1$$

where I_k is a k-dimensional identity matrix. Processes in X are now constructed using a k-dimensional vector ϕ of functions in L^2 via:

$$x(t) = \int_{-\infty}^{+\infty} \phi(t) \cdot dW(t - \tau) \, .$$

The analyses in Sections 2 and 3 extend by applying the decompositions to each of the k Laplace transforms of entries in ϕ. In Example 1 formula (3.4) still applies where $\mathcal{L}p\,(\psi)$ is the vector of Laplace transforms of entries in ψ. The recursions derived in Example 3 still apply where q_n is now a k-dimensional vector of polynomials, each with orders less than the scalar polynomial q_d.

5. Nonstationarities

In Section 3, the assumption that $\psi \in L_+^2$ guaranteed that process y is second moment stationary. Our analysis can be extended to a more general class of processes, however. To accommodate nonstationarities, it is most convenient to think of the underlying information process as starting at some initial time, say $t = 0$. Hence we imagine (1.1) and (1.2) holding for nonnegative values of t_1, t_2, t_3 and t_4, and we assume that the random measure of any interval contained in $(-\infty, 0)$ is zero. This permits formula (3.1) to be well defined for a much larger class of functions ψ. We might view the process y as being the deviation from a path that is perfectly predictable from time zero forward. We impose the weaker requirement that $\exp(-\rho t)\psi$ be in L^2 for strictly positive values of ρ which allows for polynomial growth in the second moment of y.[3] The calculations in Examples 1 through 3 still apply. In the case of Example 3, to accommodate polynomial growth we now allow q_d to have zeros on the imaginary axis of the complex plane $\mathbf{C}$.

Notes

1. The uniqueness of this decomposition requires some qualification. Elements of L^2 are only defined up to an equivalence of functions that are equal almost everywhere. Hence from the vantage point of L^2, the construction of ϕ^+ and ϕ^- at a particular point, say $t = 0$, is inconsequential.

2. We take the right side of equation (3.1) as the definition of $x(t)$. Alternatively, for particular classes of γ we could define $x(t)$ using finite sum approximations for the middle integral.

3. In Examples 1 and 3 it is also possible to allow for exponential growth in the second moments of y as long as $\psi \exp(-\sigma t)$ is in L^2_+ for some σ satisfying $0 < \sigma < \delta$. The transform analysis now applies to the narrower strip $\mathbf{C}_\delta^- \cap \mathbf{C}_\sigma^+$.

9

Identification of Continuous Time Rational Expectations Models from Discrete Time Data

by Lars Peter HANSEN and Thomas J. SARGENT

1. Introduction

This paper proves two propositions about identification in a continuous time version of a linear stochastic rational expectations model. The model is a continuous time version of Lucas and Prescott (1971), in which the equilibrium can be interpreted as the solution of a stochastic control problem, either of a collection of private agents or of a fictitious *social planner*. Estimation is directed toward isolating the parameters of the *agent's* objective function and of the stochastic processes of the forcing functions that the agent faces. This approach has been advocated by Lucas (1967, 1976), Lucas and Prescott (1971), and Lucas and Sargent (1981) as offering the potential to analyze an interesting class of policy interventions promised by *structural* models, while meeting the criticisms of most econometric policy evaluation methods that were made by Lucas (1976). At the same time, inspired by the work of Sims (1971), Geweke (1978), and P.C.B. Phillips (1972, 1973, 1974), we want to estimate models in which optimizing economic agents make decisions at finer time intervals than the interval of time between the observations used by the econometrician. We adopt a continuous time theoretical framework both because it is an interesting limiting case, and because it has received extensive attention in the theoretical and the econometric literatures.

Identification of the parameters of a continuous time model from discrete time data must confront the aliasing problem (see, e.g., Phillips 1973). In general, there is an uncountable infinity of continuous time models that are consistent with a collection of discrete time observations. However, with finite parameter continuous time models, the aliasing problem, while still present, is less severe. The dimensions of the aliasing identification problem for the particular class of finite parameter models treated in this paper have been studied in earlier papers by Phillips (1973) and Hansen and Sargent (1983b). In these finite parameter models, there is a finite number of observationally equivalent continuous time models that are consistent with the discrete time observations. To achieve identification of the continuous time model, an additional source of prior restrictions is needed. This paper shows how the non-linear cross-equation restrictions implied by rational expectations achieve identification of the continuous time model.

We consider a linear rational expectations model that gives rise to systems of stochastic differential and difference equations that resemble the forms of Phillips' (1973) systems. However, we analyze identifying restrictions of a different variety than those studied by Phillips. As Lucas (1976), Lucas and Sargent (1981), and Hansen and Sargent (1980a, 1981a, 1981d) have pointed out in several related contexts, even rational expectations models that are linear in the variables typically are characterized by sets of highly nonlinear cross-equation restrictions, which to a large extent replace the linear (usually exclusion, usually within-equation) restrictions used to identify many existing time series models.

The intuition underlying our results is as follows. In dynamic rational expectations models, agents' decisions partly depend on their expectations of all future values of other variables in the model. When agents are acting in continuous time, a discrete time record of agents' decisions contains information about their forecasts of other variables in the model for all instants in the future. Under rational expectations, the hints about agents' views of the future contained in their decisions at discrete points in time restrict the actual behavior of these other variables as stochastic processes in continuous time. These hints are the source of identification that we propose to utilize.

We prove identification propositions under two alternative sets of conditions. The first set of conditions severely restricts the serial correlations of the unobservable disturbance term, although it does not require that the right-hand-side observables be strictly exogenous. The second set of conditions leaves the serial correlations of the disturbance unrestricted but imposes that the right-hand-side variables must be strictly exogenous in continuous time and that they have a rational spectral density matrix. Identification is then achieved from the restrictions that the theory imposes between the projections of the endogenous on the exogenous variables, on the one hand, and the spectral density matrix of the exogenous variables, on the other hand. This second set of conditions thus uses an approach to identification in the spirit of that used by Hatanaka (1975) in the context of discrete time models. Our results exhibit a tradeoff between the strength of strict exogeneity and serial correlation assumptions that are sufficient for identification. A similar tradeoff occurs in discrete time series models.

2. The Continuous Time Model

The model studied is a continuous time, linear-quadratic version of a Lucas-Prescott model of investment under uncertainty. This model has a variety of possible interpretations, applications, and extensions (for example, see Hansen and Sargent 1981a, Eckstein 1984, and Eichenbaum 1983). For the identification propositions proved here, a single factor model involving a single dynamic decision variable is used. In the appendix, we briefly indicate how the results might be extended to prove identification of continuous time, interrelated factor models from discrete time data.[1]

Consider a firm or fictitious social planner that maximizes over strategies for $K(t)$ the criterion

$$E_0 \int_0^\infty J[K(t), DK(t), t, z_1(t), y(t)]dt \tag{1}$$

where

$$\begin{aligned} &J[K(t), DK(t), t, z_1(t), y(t)] \\ &= \left\{y(t)K(t) - \beta K(t)^2 - z_1(t)DK(t) - \alpha[DK(t)]^2\right\}e^{-rt}, \end{aligned}$$

where D is the time derivative operator, and where E_t is the expectations operator conditioned on information available at time period t. Here $K(t)$ is the capital stock, $z_1(t)$ is the relative price of investment goods, $y(t)$ is a random shock to productivity, all at time period t, α and β are positive constants, and r is a fixed discount rate. The variables $z_1(t)$ and $y(t)$ are elements in a vector stochastic process of forcing variables. Using results from Hansen and Sargent (1981d) and Chapter 7, the Euler equation for the certainty equivalent version of the firm's maximization problem is

$$\begin{aligned} &-\alpha D^2K(t) + r\alpha DK(t) + \beta K(t) \\ &= -(1/2)\,[rz_1(t) - y(t) - Dz_1(t)] \end{aligned} \tag{2}$$

For simplicity, we assume that the discount rate is zero.[2] The characteristic polynomial for the Euler equation (2) can be factored

$$-\alpha s^2 + \beta = (\rho - s)\ (\rho + s)\alpha$$

where

$$\rho = \sqrt{\frac{\beta}{\alpha}}\ .$$

The solution to the Euler equation (2) that maximizes (1) is

$$(3)\ \ DK(t) = -\rho K(t) - (1/2\alpha)E_t \int_0^\infty e^{-\rho\tau}[Dz_1(t+\tau) + y(t+\tau)]d\tau\ .$$

We seek to identify ρ, α, and the parameters of the stochastic processes of the forcing variables from discrete time data.[3] To provide an interpretation of the error term in equations fit by an econometrician, we assume that $y(t)$ is observed by private agents but not by the econometrician. Let $z(t)' = [z_1(t), z_2(t)']$, where $z_2(t)$ is a list of additional variables which are seen by both private agents and the econometrician and which help predict future z_1's. The econometrician knows the discrete time covariogram and cross-covariogram of the (K, z) process and from these moments seeks to identify the parameters ρ and α that characterize the continuous time objective function (1) and the parameters of the continuous time stochastic process governing (z, y). We study this identification problem using two alternative specifications of the continuous time stochastic process (z, y).

3. Identification Where (K, z) Is a First-Order Markov Process

In this section we make a special assumption about the forcing variables that is sufficient to imply that (K, z) is a covariance stationary, first-order Markov process. Specifically,

Assumption 1: The forcing variables $y(t)$ and $z_1(t)$ are governed by[4]

$$y(t) = D\varepsilon_1^*(t)$$

and

$$(4)\qquad Dz(t) = A_{22}z(t) + \varepsilon_2^*(t)$$

where $z_1(t)$ is the first element in the $n-1$ dimensional vector $z(t)$, the eigenvalues of A_{22} have negative real parts, and $\varepsilon^{*\prime} = [\varepsilon_1^*,\ \varepsilon_2^{*\prime}]$ is an n dimensional vector white noise with intensity matrix V_0^*.[5]

Note that Assumption 1 allows ε_1^* and ε_2^* to be correlated contemporaneously.

Using (4) and the results from Hansen and Sargent (1981d) and Chapter 8 to solve the prediction problem on the right side of (3), we obtain

$$(5) \quad DK(t) = -\rho K(t) - (1/2\alpha)uA_{22}[A_{22} - \rho I]^{-1}z(t) - (1/2\alpha)\varepsilon_1^*(t)$$

where u is the $n-1$ dimensional unit row vector given by $u = [1,0]$. We let $\varepsilon' = [(-1/2\alpha)\varepsilon_1^*,\ \varepsilon_2^*]$, and we stack equations (4) and (5) into the vector first order differential equation system:

$$\begin{bmatrix} DK(t) \\ Dz(t) \end{bmatrix} = \begin{bmatrix} A_{11} & A_{12} \\ A_{21} & A_{22} \end{bmatrix} \begin{bmatrix} K(t) \\ z(t) \end{bmatrix} + \begin{bmatrix} \varepsilon_1(t) \\ \varepsilon_2(t) \end{bmatrix}$$

or

$$Dx(t) = A_0\, x(t) + \varepsilon(t)\ .$$

The partitions of the A_0 matrix satisfy the restrictions

$$(6) \quad \begin{aligned} A_{11} &= -\rho \\ A_{21} &= 0 \\ A_{12} &= (-1/2\alpha)uA_{22}\,[A_{22} - \rho I]^{-1}\ . \end{aligned}$$

While the restriction on A_{21} is a zero restriction, the restrictions linking A_{11}, A_{12}, and A_{22} are highly nonlinear. Phillips (1973) considered the impact on identification of the zero restriction on A_{21}.[6] It happens that this exclusion restriction by itself is not sufficient to identify the parameters of A_{22} and A_{12}. However, we shall show that once we add the nonlinear cross-equation restrictions implied by rational expectations, it is possible to identify ρ, α, A_{22}, and, consequently, A_{12} and A_{11}.

It was shown by Phillips (1973) that the discrete time process X obtained by sampling x at the integers has a first order autoregressive representation,[7]

$$X(t) = B_0X(t-1) + \eta(t)$$

where

$$B_0 = \exp A_0$$

$$\eta(t) = \int_0^1 \exp(A_0\tau)\,\varepsilon(t-\tau)d\tau\ .$$

By virtue of the fact that ε is a continuous time white noise, it follows that η is a discrete time white noise. The parameters of B_0 are identified from knowledge of the discrete time matrix covariogram of the $X = (K, z)$ process.

We pose the following identification question: given the matrix B_0, is it possible uniquely to determine the free parameters of the matrix A_0?[8] That is, does the matrix equation

$$\exp A^* = B_0 = \exp A_0 \tag{7}$$

imply that $A^* = A_0$? We shall prove that the answer is yes. To proceed, we make the additional assumption:

Assumption 2: The eigenvalues of A_0 are distinct.

Write the spectral decomposition of A_0 as

$$A_0 = T\Lambda_0 T^{-1}$$

where Λ_0 is a diagonal matrix of eigenvalues of A_0 and T is the matrix whose columns are eigenvectors of A_0. Partition the matrices T and Λ_0 in the eigenvalue decomposition of A_0 comformably with A_0 so that

$$T = \begin{bmatrix} T_{11} & T_{12} \\ 0 & T_{22} \end{bmatrix}, \quad \Lambda_0 = \begin{bmatrix} \Lambda_1 & 0 \\ 0 & \Lambda_2 \end{bmatrix}.$$

It is readily verified that $-\rho = \Lambda_1$ and $A_{22} = T_{22}\Lambda_2 T_{22}^{-1}$, so that Λ_2 is the diagonal matrix of the eigenvalues of A_{22}. Now let the first $n - 1 - 2m$ eigenvalues of A_{22} be real and the remainder occur in complex conjugate pairs as $\lambda_{n-m} = \bar{\lambda}_{n-2m}, \ldots, \lambda_{n-1} = \bar{\lambda}_{n-1-m}$. For analytical convenience, we require

Assumption 3: The eigenvalues of A_0 do not differ by integer multiples of $2\pi i$.

Then if a matrix A^* is to satisfy (7), it must be related to A_0 by[9]

$$A^* = A_0 + 2\pi i T \begin{bmatrix} 0 & 0 & 0 \\ 0 & P & 0 \\ 0 & 0 & -P \end{bmatrix} T^{-1} \tag{8}$$

where P is any m dimensional diagonal matrix whose diagonal elements are arbitrary integers. In effect, (8) displays the class of perturbations of the complex eigenvalues of A_0 which leave the relation $B_0 = \exp A^*$ satisfied.

To show that the restrictions imposed on the model by rational expectations can be used to identify A_0 from B_0, we shall use the special nature of the perturbations of A_0 which are admissible under (8). Notice that all A^*'s that satisfy (8) must have identical matrices of eigenvectors—that is, T matrices—and can differ only in the imaginary parts of their complex eigenvalues. So the T matrix is identified, as are the real parts of the eigenvalues. Since ρ is a real eigenvalue, it is automatically identified. We shall indicate how the cross-equation restrictions imposed by rational expectations, in effect, link T, ρ, and the eigenvalues Λ_2. This will serve to establish the existence of a unique inverse of $B = \exp A^*$.

Using the partitioned inverse formula

$$T^{-1} = \begin{bmatrix} T_{11}^{-1} & -T_{11}^{-1}T_{12}T_{22}^{-1} \\ 0 & T_{22}^{-1} \end{bmatrix},$$

we obtain the version of the eigenvalue decomposition appropriate for our problem

$$A_0 = \begin{bmatrix} T_{11}\Lambda_1 T_{11}^{-1} & T_{12}\Lambda_2 T_{22}^{-1} - T_{11}\Lambda_1 T_{11}^{-1}T_{12}T_{22}^{-1} \\ 0 & T_{22}\Lambda_2 T_{22}^{-1} \end{bmatrix}.$$

It follows that

$$A_{12} = [T_{12}\Lambda_2 T_{22}^{-1} + \rho T_{12}T_{22}^{-1}]. \tag{9}$$

We use (6) and (9) to express the cross-equation restrictions implied by the model in the form

$$(-1/2\alpha)uA_{22}\,[A_{22} - \rho I]^{-1} = [T_{12}\Lambda_2 + \rho T_{12}]\,T_{22}^{-1}$$

or

$$(-1/2\alpha)uT_{22}\Lambda_2\,[\Lambda_2 - \rho I]^{-1} = T_{12}\,[\Lambda_2 + \rho I].$$

Solving for T_{12}, we obtain

$$(-1/2\alpha)uT_{22}\Lambda_2\,[\Lambda_2 - \rho I]^{-1}\,[\Lambda_2 + \rho I]^{-1} = T_{12}$$

or

$$T_{12} = \frac{-uT_{22}}{2}\ \text{diag}\left[\frac{\lambda_j}{(\lambda_j^2 - \rho^2)\alpha}\right]. \tag{10}$$

Since T_{12} and T_{22} are identified because the eigenvectors of A_0 are identified, equation (10) implies that the quantities

$$d_j = \frac{\lambda_j}{(\lambda_j^2 - \rho^2)\alpha} \tag{11}$$

can be inferred from the discrete time statistics. The question which remains is whether, given knowledge of d_j, ρ, and the real part of λ_j, we can infer α and the imaginary part of λ_j. To find the answer, first suppose λ_1 is real. Then it follows that α can be inferred from (11) for $j = 1$. Let j be some other index such that λ_j is complex and suppose that $\lambda_j^* = \lambda_j + 2\pi i p$ for some integer p that satisfies (11). Then we know that

$$\lambda_j(\lambda_j^{*2} - \rho^2) = \lambda_j^*(\lambda_j^2 - \rho^2) \,. \tag{12}$$

The value of λ_j^* distinct from λ_j that satisfies (12) is

$$\lambda_j^* = \frac{-\rho^2}{\lambda_j} \,. \tag{13}$$

Write $\lambda_j = \theta_1 + \theta_2 i$ where θ_1 and θ_2 are real with θ_1 less than zero. Equation (13) implies that

$$\theta_1 + (\theta_2 + 2\pi p)i = \frac{-\rho^2}{\theta_1 + \theta_2 i}$$

or

$$\begin{aligned} \theta_1\theta_2 + \pi p \theta_1 &= 0 \\ \theta_1^2 - \theta_2^2 - 2\pi p \theta_2 &= -\rho^2 \,. \end{aligned} \tag{14}$$

However, there are no values of $\{\theta_1, \theta_2, p\}$ with $\theta_1 < 0$ that satisfy both equations in (14). It follows that all of the parameters of the model are identifiable from discrete time data whenever there is at least one real eigenvalue of A_{22}.

Thus we have the following:

Proposition 1: Suppose Assumptions 1–3 are satisfied. If A_{22} has at least one real eigenvalue, then the parameters α and β (or, equivalently, α and ρ) and the parameters of A_{22} are identifiable from discrete time observations.

If there are only complex eigenvalues of A_{22}, then it can be proved, except for singular cases, that the free parameters of the continuous time model are identifiable.[10]

4. Identification with z Strictly Exogenous with Respect to K in Continuous Time

In the preceding section, the unobservable forcing variable $y(t)$ was allowed to be correlated contemporaneously with the observable forcing variables in $z(t)$. However, identification of the feedback parameter ρ used the fact that the disturbance term to the decision rule was known to be a white noise. We now wish to relax this assumption together with the assumption that the observable forcing variables can be represented as a first-order Markov process. We relax these assumptions at the cost of imposing a stronger condition about the covariance of y and z. [11]

Assumption 4: The joint process (y, z) is covariance stationary, linearly regular and satisfies the extensive orthogonality conditions $Ey(t)z(t-\tau) = 0$ for all τ.[12]

A fundamental moving average representation for (y, z) can be written in partitioned form

$$\begin{bmatrix} y(t) \\ z(t) \end{bmatrix} = \begin{bmatrix} C_1(D) & 0 \\ 0 & C_2(D) \end{bmatrix} \begin{bmatrix} \varepsilon_1(t) \\ \varepsilon_2(t) \end{bmatrix} .$$

where $C_j(s)$ is the Laplace transform of a square integrable matrix function that is concentrated on the nonnegative numbers, and where $[\varepsilon_1, \varepsilon_2']'$ is a vector white noise with intensity matrix I. For the representation to be fundamental, we must require that $[\varepsilon_1(t), \varepsilon_2(t)']'$ lie in the space spanned by linear combinations of $\{y(\tau), z(\tau);\ \tau \leq t\}$.[13]

In order to use convenient results from linear prediction theory for continuous time processes, we assume that best linear predictions and conditional expectations coincide. The forecasting problem on the right side of equation (3) can be solved using techniques developed in Chapter 8 to obtain,

$$DK(t) = -\rho K(t) - \frac{u}{2\alpha}\left[\frac{DC_2(D) - \rho C_2(\rho)}{D-\rho}\right]\varepsilon_2(t) - \frac{1}{2\alpha}\left[\frac{C_1(D) - C_1(\rho)}{D-\rho}\right]\varepsilon_1(t) .$$

Next we solve for $K(t)$ and determine that[14]

(16)

$$K(t) = \frac{-u[DC_2(D) - \rho C_2(\rho)]}{2\alpha(D+\rho)(D-\rho)}\varepsilon_2(t) - \frac{1}{2\alpha}\left[\frac{C_1(D) - C_1(\rho)}{(D+\rho)(D-\rho)}\right]\varepsilon_1(t) .$$

Since $\varepsilon_1(t)$ is orthogonal to $\varepsilon_2(s)$ for all t and s, equations (15) and (16) can readily be used to calculate the projection of $K(t)$ onto current, past, and future z's. This projection is given by

$$K(t) = \frac{-u[DC_2(D) - \rho C_2(\rho)]}{2\alpha(D+\rho)(D-\rho)} C_2(D)^{-1} z(t) + \xi(t)$$
$$= \frac{-u[DI - \rho C_2(\rho) C_2(D)^{-1}]}{2\alpha(D+\rho)(D-\rho)} z(t) + \xi(t)$$

where[15]

$$Ez(t)\xi(t-\tau) = 0 \text{ for all } \tau \, .$$

It is instructive to calculate the discrete time cross spectral density of K and z:

$$F_1(\omega) = \sum_{j=-\infty}^{+\infty} \left\{ -u \left[\frac{i(\omega + 2\pi j) C_2(i\omega + 2\pi i j) - \rho C_2(\rho)}{2\alpha[-(\omega+2\pi j)^2 - \rho^2]} \right] C_2(-i\omega - 2\pi i j)' \right\} . \tag{17}$$

From discrete time data we can identify the function F_1 together with the discrete spectral density of z which is given by

$$F_2(\omega) = \sum_{j=-\infty}^{+\infty} C_2(i\omega + 2\pi i j) C_2(-i\omega - 2\pi i j)' \, . \tag{18}$$

The cross-equation rational expectations restrictions are apparent in that the parameterization C_2 occurs in both the spectral density matrix F_2 and in the cross spectral density F_1. The identification question is whether the function C_2 and the parameters ρ and α can be inferred from F_1 and F_2 using relations (17) and (18). Without imposing additional restrictions on C_2, the answer to this question would appear to be no. However, once we restrict the admissible parameterizations of C_2 to be rational in the way described by Hansen and Sargent (1983b), we can achieve identification.

To achieve identification, we impose the following additional assumption. Define $\lambda_0 = -\rho$.

Assumption 5: $C_2(s)$ is of the form

$$C_2(s) = \frac{G_0 + G_1 s + \ldots G_{q-1} s^{q-1}}{(s-\lambda_1)(s-\lambda_2) \, \ldots \, (s-\lambda_q)}$$

where G_0, G_1, $\ldots G_{q-1}$ are $(n-1) \times (n-1)$ real matrices. The zeros of det $G(s)$ have negative real parts and $G(\rho) + G(-\rho)$ is nonsingular where $G(s) = G_0 + G_1 s + \ldots + G_{q-1} s^{q-1}$. [16] In addition, we assume that $G(\lambda_j) \neq 0$ and the real part of λ_j is strictly negative for $j = 1, 2, \ldots, q$. Finally, for each j (including $j = 0$), $\lambda_j = \bar{\lambda}_k$ for some index k, and any two λ's with the same real part do not have imaginary parts that differ by integer multiples of $2\pi i$.

The λ's (for $j = 1, 2, \ldots, q$) are called the poles of $C_2(s)$.

With this specification for $C_2(s)$, the spectral density of z is known to have the form

$$f_2(\omega) = \sum_{j=1}^{q} \left[\frac{Q_j}{i\omega - \lambda_j} + \frac{Q_j'}{-i\omega - \lambda_j} \right] , \tag{19}$$

where

$$Q_j = \lim_{s \to \lambda_j} (s - \lambda_j) C_2(s) C_2(-s)' ,$$

f_2 is the spectral density matrix of z, and the prime denotes transposition but not conjugation. See Hansen and Sargent (1983b) or A.W. Phillips (1959) for the details of this construction. From (16) we can deduce that the cross spectral density matrix is rational.[17] In particular, let

$$h_1(s) = \frac{-u[sC_2(s) - \rho C_2(\rho)] C_2(-s)'}{2\alpha(s^2 - \rho^2)} .$$

Then the cross spectral density of z and K is given by $f_1(\omega) = h_1(i\omega)$. We form a partial fractions representation of h_1 to obtain

$$h_1(s) = \sum_{j=0}^{q} \left[\frac{\hat{Q}_j}{s - \lambda_j} + \frac{\tilde{Q}_j}{-s - \lambda_j} \right]$$

where

$$\hat{Q}_j = \lim_{s \to \lambda_j} (s - \lambda_j) h_1(s)$$

and

$$\tilde{Q}_j = - \lim_{s \to -\lambda_j} (s + \lambda_j) h_1(s) .$$

Note that for $j = 1, 2, \ldots, q$

$$\hat{Q}_j = \frac{-u \lambda_j Q_j}{2\alpha(\lambda_j - \rho^2)} , \tag{20}$$

and

$$(21) \qquad \hat{Q}_0 = \frac{u[-\rho C_2(-\rho) - \rho C_2(\rho)]\, C_2(\rho)'}{4\alpha\rho} \,.$$

Since $\hat{Q}_0$ is different from zero, ρ can be identified from the discrete-time cross spectral density F_1. To identify α and the imaginary parts of the poles of $C_2(s)$, we make use of the fact that Q_j, $\hat{Q}_j$, and the real parts of the poles of $C_2(s)$ are identifiable from discrete time data and that (20) holds. The matrices Q_j and $\hat{Q}_j$ can be inferred from the discrete-time spectral density of z and the cross spectral density of K and z, respectively. The real parts of the poles of $C_2(s)$ can be inferred from the discrete time spectral density F_2 of z (see Phillips 1959 and Hansen and Sargent 1983b). Equation (20) is a restriction across the spectral density of z and the cross spectral density of K and z. Using (19) and (20) we see that the quantities

$$(22) \qquad d_j = \frac{\lambda_j}{\alpha(\lambda_j^2 - \rho^2)}$$

are identified from discrete time statistics. Equation (21) is identical with equation (11) derived for the first-order Markov case. We summarize these results in

Proposition 2: Suppose Assumptions 4 and 5 are satisfied. If there is at least one real pole of $C_2(s)$, then the parameters α and β (or, equivalently, α and ρ) and the continuous time spectral density matrix of z are identifiable from discrete time observations.

If there fail to be any real poles of C_2, then all of the parameters can still be identified except possibly for some singular cases.

5. Conclusions

The two propositions proved in this paper indicate how the cross-equation restrictions of rational expectations models can serve to identify the parameters of a continuous time model from discrete time observations. The basic idea is that where decisions reflect forecasting in continuous time, the discrete time data on the decision variable and the forcing variables contain adequate clues to permit us to infer the parameters of the joint continuous time process of decision and forcing variables.

The basic identification mechanism promises to carry over to more complicated specifications than the two that are formally analyzed in this paper. Extensions to our two specifications can be imagined in a variety of directions. These include

- Higher order Markov schemes for the forcing process $z(t)$ in our first setup.
- Higher order processes for the unobservable $y(t)$ in our first setup.
- Multiple interrelated decision variables.
- Higher order adjustment costs.

A formula expressing the cross-equation restrictions for a multiple decision variable problem that is highly suggestive of identification, though falling short of providing a formal proof, is reported in the appendix.

This paper is intended as a prologue to Hansen and Sargent (1980b) that describes methods for estimating continuous time linear rational expectations models that generalize the models analyzed in this present paper. While formal identification theorems are not yet available for those more general models, a method of checking for the presence of an aliasing identification problem is readily available in any particular application.[18]

Appendix

In this appendix we consider a multiple decision variable version of the quadratic optimization problem considered in Section 2. We let $K(t)$ be a p dimensional decision vector, $z_1(t)$ be a p dimensional vector of forcing variables that are observed by the econometrician, and $y(t)$ be a p dimensional vector of forcing variables that are not observed by the econometrician. We consider a firm that maximizes over strategies for $K(t)$ the criterion

$$E_0 \int_0^\infty J[K(t), DK(t), t, z_1(t), y(t)]dt$$

where

$$\begin{aligned} &J[K(t), DK(t), t, z_1(t), y(t)] \\ &\quad = \{y(t)'K(t) - K(t)'\beta K(t) - z_1(t)'DK(t) \\ &\quad - [DK(t)]'\alpha[DK(t)]\}e^{-rt} . \end{aligned}$$

Here α and β are $p \times p$ positive definite matrices. We assume that

$$y(t) = D\varepsilon_1^*(t)$$

and

$$Dz(t) = A_{22}z(t) + \varepsilon_2^*(t) \tag{23}$$

where $z_1(t)$ is a p dimensional subvector of the $n-p$ dimensional vector $z(t)$, the eigenvalues of A_{22} have negative real parts, and $\varepsilon^{*\prime} = [\varepsilon_1^{*\prime}, \varepsilon_2^{*\prime}]$ is an n dimensional vector white noise with intensity matrix V_0^*. We factor the characteristic polynomial of the Euler equation

$$-\alpha s^2 + (r^2/4)\alpha + \beta = [a - bs]'[a + bs]$$

where a and b are each $p \times p$ matrices such that the zeros of $\det(a - bs)$ lie in the right-half plane while the zeros of $\det(a + bs)$ lie in the left-half plane. This factorization is unique up to a premultiplication of a and b by a common orthogonal matrix.

Using results in Hansen and Sargent (1981d) and in Chapter 7, we find that the solution to the maximization problem of the firm is

$$\begin{aligned} DK(t) = -[b^{-1}a - \frac{r}{2}I]K(t) + \frac{1}{2}\sum_{j=1}^{p} \Big\{ N_j u[A_{22} - rI] \\ [A_{22} - (s_j + \frac{r}{2})I]^{-1} z(t) \Big\} - \frac{1}{2}\sum_{j=1}^{p} N_j \varepsilon_1^*(t) \end{aligned} \tag{24}$$

where

$$\det[a'b - sb'b] = s_0(s - s_1) \ \ldots \ (s - s_m)\,,$$

$$N_j = \frac{\operatorname{adj}[a'b - b'bs_j]}{s_0 \Pi_{i \neq j}(s_i - s_j)}\,,$$

and u is a $p \times (n-p)$ matrix of the form $u = [I, 0]$. We can write (23) and (24) as the joint first-order differential equation

$$Dx(t) = A_0\, x(t) + \varepsilon(t)$$

where

$$\begin{aligned} A_0 &= \begin{bmatrix} A_{11} & A_{12} \\ A_{21} & A_{22} \end{bmatrix} \\ A_{11} &= -[b^{-1}a - \frac{r}{2}I] \\ A_{12} &= \frac{1}{2}\sum_{j=1}^{p} N_j u[A_{22} - rI]\,[A_{22} - (s_j + \frac{r}{2})I]^{-1} \\ A_{21} &= 0 \\ x(t) &= \begin{bmatrix} K(t) \\ z(t) \end{bmatrix} \qquad \varepsilon(t) = \begin{bmatrix} -\frac{1}{2}\sum_{j=1}^{p} N_j \varepsilon_1^*(t) \\ \varepsilon_2^*(t) \end{bmatrix}. \end{aligned} \tag{25}$$

As in the third section, we ask whether the matrix equation

$$\exp A^* = B_0 = \exp A_0$$

implies that $A^* = A_0$. Assume that the eigenvalues of A_0 are distinct and that they do not differ by integer multiples of $2\pi i$. Write the spectral decomposition of A_0 as $T\Lambda_0 T^{-1}$ where Λ_0 is the diagonal matrix of eigenvalues and T is a matrix whose columns are eigenvectors of A_0. Partition the matrices T and Λ_0 conformably with A_0 so that

$$T = \begin{bmatrix} T_{11} & T_{12} \\ 0 & T_{22} \end{bmatrix} \qquad \Lambda_0 = \begin{bmatrix} \Lambda_1 & 0 \\ 0 & \Lambda_2 \end{bmatrix} .$$

It follows that

$$A_{12} = T_{12}\Lambda_2 T_{22}^{-1} - A_{11}T_{12}T_{22}^{-1} .$$

Restriction (25) implies that

$$(26)\quad T_{12}\Lambda_2 - A_{11}T_{12} = \frac{1}{2}\sum_{j=1}^{p} N_j u[A_{22} - rI]\,[A_{22} - (s_j + \frac{r}{2})I]^{-1}T_{22} .$$

Let vec$(\cdot)$ represent the vector formed by taking the direct sums of the rows of a matrix, and let $\otimes$ denote the Kronecker product. We solve (26) for T_{12} to obtain

$$(27)\qquad \text{vec}\ T_{12} = [(-A_{11} \otimes I) + (I \otimes \Lambda_2)]^{-1}\ \text{vec}\ c$$

where

$$c = \frac{1}{2}\sum_{j=1}^{p} N_j u[A_{22} - rI]\,[A_{22} - (s_j + \frac{r}{2})I]^{-1}T_{22} .$$

From our discussion in the third section, we know that the eigenvector matrix T and the real parts of the eigenvalues in Λ_0 can be inferred from discrete time data. The imaginary parts of the complex eigenvalues can be perturbed by adding integer multiples of $2\pi i$ such that the complex conjugate pairs remain intact to generate alternative choices of A^* that satisfy

$$\exp A^* = B_0 .$$

However, (27) restricts the class of *admissible perturbations* of the eigenvalues further so that it appears that in most circumstances A_0 is identifiable from discrete time data as are α and β.

Notes

1. This class of models includes continuous time, linear stochastic versions of the models discussed by Gould (1968), Lucas (1967), Mortensen (1973), and Treadway (1969). Geweke (1977a) used a model of this kind to motivate interpretations of some discrete time regressions.

2. Our discussion could be modified in a straightforward way to accommodate situations in which r is specified *a priori* but is different from zero. When r is set to zero, we have to interpret the decision rule we investigate as a limit of decisions rules as r declines to zero.

3. Given that ρ and α are identified, β can be inferred from the relation $\beta = \rho^2/\alpha$.

4. The assumption that y is the derivative of the white noise ε_1^* is contrived to imply that the decision rule has a white noise disturbance. In our discussion, the means of all of the random variables have been implicitly set to zero.

5. For an introduction to continuous time, linear stochastic processes, see Kwakernaak and Sivan (1972). A continuous time vector white noise $\varepsilon(t)$ is said to have intensity matrix V if $E\varepsilon(t)\varepsilon(t-\tau) = \delta(\tau)V$ where δ is the Dirac delta generalized function.

6. Phillips (1973) has also studied cross-equation linear restrictions.

7. See Kwakernaak and Sivan (1972), Coddington and Levinson (1955), and Gantmacher (1959) for the definition and properties of the matrix exponential function $\exp(A)$.

8. Hansen and Sargent (1983b) showed that there is extra identifying information about A_0 contained in the expression linking the covariance matrix of η to the intensity matrix of ε. In our discussion below, we supply sufficient conditions for identification that do not exploit this extra information.

9. See Coddington and Levinson (1955) or Gantmacher (1959).

10. For example, if there is only one complex conjugate pair of eigenvalues of A_{22} and no real eigenvalues, then it can be shown that the imaginary part of one of these eigenvalues has to satisfy a cubic equation. Unless the cubic equation has solutions that differ

by an integer multiple of 2π, identification of the continuous time parameters is achieved. Thus, identification will only be a problem in singular cases. The existence of multiple pairs of complex conjugate eigenvalues of A_{22} will make identification even less likely to be a problem.

11. For discrete time models, Hatanaka (1975) treated the identification of structural parameters from the projections of the endogenous on the exogenous variables without using prior information about the orders of serial correlation of disturbance processes.

12. See Rozanov (1967) for a definition of the term linearly regular.

13. See Hansen and Sargent (1983b), for a fuller technical description of the setup being used here.

14. Here we have implicitly assumed that the decision rule of the firm has been employed forever.

15. Here we have implicitly assumed that z has a continuous time autoregressive representation. We do not need to make this assumption in what follows.

16. This is one of the setups used by Hansen and Sargent (1983b). They provide more technical details.

17. Although the spectral density of z and the cross spectral density of K and z are rational, the spectral density of K is not necessarily rational and is not necessarily identifiable from discrete time data.

18. The method involves calculating the poles of the estimated stochastic process of the forcing variables and constructing an observationally equivalent continuous time model by perturbing the complex eigenvalues by integer multiples of $2\pi i$. It can then be checked whether the implied continuous time model for the joint process of decision variables and forcing variables is observationally equivalent with the estimated model.

10

Temporal Aggregation of Economic Time Series

by Albert MARCET

Introduction

We call temporal aggregation the situation in which a variable that evolves through time can not be observed at all dates. This phenomenon arises frequently in economics, where it is very expensive to collect data on certain variables, and there is no reason to believe that economic time series are collected at the frequency required to fully capture the movements of the economy. For example, we only have quarterly observations on GNP, but it is reasonable to believe that the behavior of GNP within a quarter carries relevant information about the structure of the economy.

In order to give a mathematical structure to this problem, we assume that there is an underlying stochastic process in continuous time that is observed only at discrete intervals. This structure has been used by Sims (1971) and Geweke (1978) to describe the effects of temporal aggregation in the distributed lag model.

We will be concerned with the issues that arise in the study of linear predictions of future values of the variables given all information up to the present. In other words, if we have a vector of n variables y, we try to predict $y_i(t+1)$ using a function of the form $\sum_{k=0}^{\infty} \mu_k' \, y(t-k)$.

Since the fundamental *moving average representation* (henceforth MAR) of a stochastic process (or Wold decomposition) is a good summary of the properties of those predictions, we will describe what is the relationship between the Wold decomposition of the unobserved continuous time process and the Wold decomposition of the discrete sampled process.

This approach to time series analysis has been used widely in economics. In rational expectations models, agents are often assumed to form their expectations using the kind of predictions mentioned above and the MAR has proved to be a useful tool in solving these models. Also, Sims (1980) has developed a method for interpreting time series using the MAR; this approach has been used in Litterman and Weiss (1985) and Bernanke (1986).

We focus on the following type of questions. What features of the continuous MAR are captured by the discrete MAR? What type of systematic biases will exist in the discrete MAR? How are the one-step-ahead prediction errors in continuous and discrete time related? Can we approximate in any sense the continuous model arbitrarily well by sampling the continuous process more frequently?

Our study can be useful in deciding if the results obtained in a given application of time series analysis can be attributed to time aggregation bias. If the econometrician suspects that this is the case, he should look for data collected at a finer interval or, alternatively, he could estimate a structural model in continuous time with discrete sampled data using, for example the techniques developed by Hansen and Sargent (1980b) and (1981d). These methods have been applied in Hansen and Sargent (1983a), Christiano (1984) and Christiano and Eichenbaum (1985, 1986). Other techniques have also been developed by Bergstrom (1976, 1983) and by Lo (1985).

In order to analyze the relationship between the continuous and discrete MAR, we introduce an approach that relates these from the point of view of the space of functions L^2. As an application of this approach, we analyze the effects of using data that have been averaged over a certain period of time, and show that a systematic bias will be present when this type of data are mixed with sampled data.

Due to the nature of the subject, it is inevitable that the exposition will become technical. An effort has been made in order to clarify the meaning of the propositions, and to give them an intuitive interpretation. In the conclusion, the main results of the paper are restated in a simple way, and some direct implications of our work for econometric practice are briefly discussed.

In terms of the issues addressed, our work is in the same spirit as that of Sims (1971) and Geweke (1978). It is worth noting that the distributed lag model (studied by Sims and Geweke), and the study of predictions using all data up to the present (studied in this paper) are two basically different ways of analyzing time series, so that our work is not an extension or a particular case of their model. Interestingly enough, in the distributed lag model, one effect of temporal aggregation is that the first few coefficients of the discrete approximation will, in general, be small compared with the corresponding coefficients in continuous time, while the effect of temporal aggregation on the MAR is, in general, the oppposite (see Section 3).

We will be concerned with a very general class of n-dimensional stochastic processes to be specified in Section 1, where we set up our problem and our notation. In Section 2, we discuss the relationship between the discrete and the continuous one-step-ahead innovations. In Section 3, we display a formula relating the MAR of the continuous process, to the MAR of the discrete (sampled) process, and we compare it with the formulas obtained by Sims and Geweke in the distributed lag model. We find that there is an intimate relationship between how good the predictions with sampled data are, and how well the discrete MAR approximates the continuous MAR; also, we use these theoretical results to discuss how and when the shape of the MAR of the sampled process will be very different from the shape of the continuous MAR. Section 4 contains certain results and examples that illustrate how Granger causality relationships are affected by time aggregation. The effects of using unit-averaged data, are studied in Section 5. In Section 6, we obtain some preliminary results on the issue of convergence of the MAR as the sampling frequency increases; the results in this section analyze the approximation of the continuous projection with the discrete projection. Finally, we discuss in an appendix the autoregressive representation of the sampled process; the results in the appendix are useful in other parts of the paper.

1. Definition of the Problem

Let us consider an n-dimensional stochastic process in continuous time y; we will denote by $H_y(t)$ the set that contains all linear combinations of $y_i(s)$ for all $s \leq t$ and $i = 1, \ldots, n$, plus the limits of convergent sequences of these linear combinations. More formally, $H_y(t) = cl\{\text{linear subspace generated by } y_i(s), \text{ for all } s \leq t,\ i = 1, \ldots, n\}$, (where the closure is taken with respect to a metric to be specified below). Intuitively, we can think of this set as containing all random variables of the form

$$\sum_{k=0}^{\infty} \mu_k' y(t_k)$$

where $t_k \leq t$, and $\mu_k \in \mathbf{R}^n$ are constant vectors.

We will treat the space of random variables as a metric space with the distance given by the mean-square difference; i.e., for any two random variables x, z,

$$d(x, z) = (E(x - z)^2)^{\frac{1}{2}} .$$

One of the implications of using this distance, is that two random variables x, z will be considered equal when they are equal with probability one.

Consider the problem of predicting $y(t+\alpha)$, $\alpha > 0$, using all the information available up to time t. We define the best linear predictor of $y(t + \alpha)$ with information up to t, as the element of $H_y(t)$ that minimizes the mean square error

$$E(y(t+\alpha) - x)^2 \qquad \text{for } x \in H_y(t) .$$

We will call this predictor the projection of $y(t + \alpha)$ on $H_y(t)$, and we will denote it by $\eta(y(t + \alpha)|H_y(t))$. More formally, we define $\eta(y(t + \alpha)|H_y(t))$ as the element of $H_y(t)$ such that

$$d\Big(\eta(y(t+\alpha)|H_y(t)),\, y(t+\alpha)\Big) \leq d\Big(x,\, y(t+\alpha)\Big) \quad \text{for any} \quad x \in H_y(t) .$$

Next, we are going to define the problem of temporal aggregation.

Let (Ω, S, P) be a probability space, and let y be a multivariate stochastic process in continuous time, $y : \mathbf{R} \times \Omega \to \mathbf{R}^n$. In order to guarantee that a moving average representation exists in continuous time, we make the following assumptions on this process:

1. $E(y_i(t)) = 0$, $\text{var}(y_i(t)) < \infty$ for all $i = 1, \ldots, n$, and all $t \in \mathbf{R}$
2. y is covariance stationary.
3. y is linearly regular.

Then, letting $m \leq n$ be the rank of the process y, this process has a fundamental (or Wold) moving average representation (MAR), which can be expressed by

$$y(t) = \int_0^\infty a(u)\, \zeta\,(t - du) \tag{1}$$

where $a : \mathbf{R} \to \mathbf{R}^{n \times m}$, and ζ is a vector of m orthonormal random measures. The matrix function a is unique up to multiplication by an orthonormal matrix.[1]

In this paper, we will often use the following properties of random measures: for any two square integrable functions $f, g : \mathbf{R} \to \mathbf{R}^{n \times m}$, the integrals $\int f d\zeta$ and $\int g d\zeta$ are well defined random variables, and they satisfy: $E(\int f d\zeta \cdot \int g' d\zeta) = \int f \cdot g'$. We can use this property to find the covariance function of y; since

$$E\left(y(t) \cdot y(t-s)'\right) = \int_0^\infty a(u)a(u-s)'du \quad \text{all } \ s \in \mathbf{R}\,.$$

In the rest of the paper, the initials MAR will stand for the fundamental moving average representation.

An important property of the MAR is that

$$\eta(y(t)|H_y(t-\alpha)) = \int_\alpha^\infty a(u)\zeta(t-du) \qquad \text{for all } \alpha > 0\,.$$

Using the process y, we define another stochastic process in discrete time $Y : I \times \Omega \to \mathbf{R}^n$ on the same probability space (Ω, S, P), by setting

$$Y(t, \omega) = y(t, \omega) \qquad \text{for all } \ t \in I, \ \text{all } \omega \in \Omega \tag{2}$$

where I is the set of all integers. We will call Y the sampled process. In our notation, we will suppress the dependence of Y on ω.

Most of this paper relates the MAR of y to the MAR of Y, so that we have to guarantee that the MAR of Y exists.

Indeed, we can see that Y inherits all the 'good' properties of y. By definition, Y satisfies assumptions 1 and 2. We show next that Y is also linearly regular.

Let $\{x_k\}$ be any sequence of random variables. By definition of mean square convergence, $x_k \to 0$ as $k \to \infty$ if and only if $\text{var}(x_k) \to 0$ as $k \to \infty$. Since for each integer s $H_Y(s) \subset H_y(s)$, we have

$$\begin{aligned} 0 \le \text{var}\Big(\eta(Y_i(t)|H_Y(s))\Big) &= \text{var}\Big(\eta(y_i(t)|H_Y(s))\Big) \\ &\le \text{var}\Big(\eta(y_i(t)|H_y(s))\Big), \qquad\qquad i = 1, \ldots, n\,. \end{aligned}$$

Since y is assumed to be linearly regular, the term to the right of this string of inequalities goes to zero as $s \to -\infty$, so that the variance of $\eta(Y_i(t)|H_Y(s))$ also goes to zero, and Y is linearly regular.

Hence, the MAR of Y exists, and we can write

$$Y(t) = \sum_{k=0}^{\infty} A_k\, \epsilon(t-k) \tag{3}$$

where A_k are $n \times m'$ matrices, and ϵ is an m'-dimensional vector of white noises with

$$\begin{aligned} E(\epsilon(t) \cdot \epsilon'(s)) &= 0 \quad \text{if } \ t \ne s \\ &= \Sigma \quad \text{if } \ t = s\,. \end{aligned}$$

Note that, unlike in the continuous time case, we do not require that the elements of ϵ be uncorrelated contemporaneously.

The requirement that (3) is the fundamental moving average representation for Y guarantees that

$$\eta(Y(t)|H_Y(t-r)) = \sum_{k=r}^{\infty} A_k\, \epsilon(t-k) \quad \text{for any integer } r \geq 0\,.$$

We will say that ζ and ϵ are fundamental for y and Y, respectively; it can be shown that $H_y(t) = H_\zeta(t)$ and $H_Y(t) = H_\epsilon(t)$.

It is possible to normalize the discrete MAR so that the A_k's and ϵ are uniquely determined. In our study, different normalizations will be useful for different purposes, and we leave this question open for the moment.

We can extend the function a to the negative reals by setting $a(u) = 0$ if $u < 0$ and, analogously, we set $A_k = 0$ for any integer $k < 0$. After doing this, we can express (1) and (3) in convolution notation:

$$\begin{aligned} y(t) &= a * \zeta(t) \qquad t \in \mathbf{R} \\ Y(t) &= A * \epsilon(t) \qquad t \in \mathbf{I}\,. \end{aligned}$$

2. The Innovation of the Discrete Sampled Process

Throughout this and the next two sections, we normalize the discrete MAR by setting $\epsilon_i(t) = Y_i(t) - \eta(Y_i(t)|H_Y(t-1))$, $i = 1, \ldots, n$; that is, $\epsilon(t)$ is the vector of one-step-ahead innovations in Y.

As Hansen and Sargent (1984, reprinted as Chapter 4 of this volume) point out, the innovation in discrete time can be written as

$$\epsilon(t) = \int_0^1 a(u)\zeta(t-du) + B_t \tag{4}$$

where $B_t = \eta(y(t)|H_y(t-1)) - \eta(Y(t)|H_Y(t-1))$. Clearly, since $H_Y(t-1) \subset H_y(t-1)$, we have $B_t \in H_y(t-1) = H_\zeta(t-1)$, so that $B_t \perp \int_0^1 a(u)\zeta(t-du)$. In words $\epsilon(t)$ is composed of two orthogonal parts: the one-step-ahead innovation in continuous time, and the difference between the one-step-ahead projections in continuous and in discrete time. By applying the law of iterated projections we have that

$$\eta(Y(t)|H_Y(t-1)) = \eta\left\{\eta(Y(t)|H_y(t-1))|H_Y(t-1)\right\}\,,$$

so that B_t could be interpreted as the error made when we try to predict the *projection* in continuous time with sampled data.

From now on, we will assume that y has full rank. This is equivalent to assuming that the spectral density of y is positive definite at almost every frequency. This guarantees that Y has full rank as well, because the spectral density of Y is given by

$$f_Y(\omega) = \sum_{k=-\infty}^{\infty} f_y(\omega + 2\pi k) \tag{5}$$

(where f_χ denotes the spectral density of χ), so that f_Y is positive definite almost everywhere.[2] Hence, m' in the MAR of Y is equal to n.

Since $B_t \in H_\zeta(t-1)$, there exists a function $c\colon [1, \infty) \to \mathbf{R}^{n\times n}$ such that[3]

$$B_t = \int_1^\infty c(u)\zeta(t - du) \, .$$

If we let $c(u) = a(u)$ for $0 \leq u < 1$, equation (4) tells us that ϵ is related to ζ by:

$$\epsilon(t) = \int_0^\infty c(u)\zeta(t - du) \, . \tag{6}$$

Hansen and Sargent (1984) point out that one implication of (6) is that the one-step-ahead prediction error $\epsilon(t)$ will in general be correlated with innovations in continuous time that have happened before $t-1$. Also, they characterize c for the case that Y has an autoregressive representation.

Next, we give a general characterization of c in terms of a which will be useful later on in the paper. First of all, observe that using the same line of argument that led us to equation (6), we can conclude that there exists a function h such that

$$\eta(Y(t)|H_Y(t-1)) = \int_1^\infty h(u)\zeta(t - du) \tag{7}$$

where $h : \mathbf{R} \to \mathbf{R}^{n\times n}$. Clearly, by the definition of h and c, we conclude that $c = a - h$. We will now characterize h.

Define the space of functions L^2_n

$$L^2_n = \left\{ f : \mathbf{R} \to \mathbf{R}^n ; \int_{-\infty}^\infty ||f(u)||^2 du < \infty \right\}$$

where, the sign $||\cdot||$ inside the integral refers to the Eucledian norm in $\mathbf{R}^n$. We will adopt the convention that each $f(u)$ is a row vector. We endow the space L^2_n with the inner product

$$\langle f, g\rangle = \int_{-\infty}^\infty f(u)\cdot g'(u)\, du = \int_{-\infty}^\infty \left[\sum_{i=1}^n f_i(u) g_i(u)\right] du \, .$$

It can be shown that this is a legitimate inner product, and that L_n^2 is a complete metric space in the metric induced by this inner product. Thus, L_n^2 is a Hilbert space.

Let us call a_i the ith row of a, so that $a_i : \mathbf{R} \to \mathbf{R}^n$, and $a(u) = \begin{bmatrix} a_1(u) \\ \vdots \\ a_n(u) \end{bmatrix}$. Clearly, each a_i belongs to L_n^2, since:

$$||a_i||^2 = \int_{-\infty}^{\infty} ||a_i(u)||^2 du = \text{var}\ (y_i(t)) < \infty \, .$$

The following proposition displays a relationship between a and h in terms of the space L_n^2.

Define the set $A \subset L_n^2$ as follows:

$$A = \text{cl}\left\{ f \in L_n^2 : f(u) = \sum_{k=1}^{s} \sum_{j=1}^{n} \mu_k^j \, a_j\,(u-k) \text{ for some finite } s \right.$$
$$\left. \text{and some } \mu_k^j \in \mathbf{R} \right\} .$$

In words, A contains all finite linear combinations of the functions $a_j(u-k)$ for $j = 1, \ldots, n$ and $k = 1, 2, \ldots$, and the limits of convergent sequences of such linear combinations.

In the first section we defined the projection of a random variable on a certain set of random variables. We can now think of doing the same with functions in L_n^2 : for a given $f \in L_n^2$ and a set $S \subset L_n^2$, we define the projection of f on S denoted by $\eta\,(f|S)$ as the element of S such that

$$d\big(\eta\,(f|S),\, f\big) \leq d\,(g,\, f) \qquad \text{for any} \quad g \in S$$

in the distance induced by the inner product of L_n^2.

<u>Proposition 1:</u> For each $i = 1, \ldots n$,[4]

$$h_i = \eta\,(a_i|A) \qquad \text{in the metric of } L_n^2 \, .$$

<u>Proof:</u> Fix i. Since A is a closed linear subspace of a Hilbert space, the projection $\eta\,(a_i|A)$ exists and it is the only element of A for which the following orthogonality conditions hold:

$$\Big[a_i - \eta\,(a_i|A)\Big] \perp f \text{ for any } f \in A. \quad \text{But}$$

$$\int_0^\infty [a_i(u) - h_i(u)] \cdot [a_j(u-k)]'du = \text{cov}\left[\int_0^\infty [a_i(u) - h_i(u)]\zeta(t-du)\,,\right.$$

$$\left.\int_0^\infty a_j(u-k)\zeta(t-du)\right] = \text{cov}\left[Y_i(t) - \eta(Y_i(t)|H_Y(t-1))\,,\, Y_i(t)\right] = 0$$

for all $k = 1, 2, \ldots, j = 1, \ldots, n$. Therefore, the orthogonality conditions in L_n^2 are satisfied with $\eta(a_i|A) = h_i$, and the only thing that is left to show is that $h_i \in A$.

By definition of $\eta(Y_i(t)|H_Y(t-1))$, there exists a sequence $\{z_\nu\}$ of random variables such that $z_\nu = \sum_{k=1}^{s_\nu} \sum_{j=1}^{n} \lambda_{k,j}^\nu Y_j(t-k)$ for some coefficients $\lambda_{k,j}^\nu$ and a finite s_ν, and such that $\{z_\nu\}$ satisfies

$$\text{var}\left[\eta(Y_i(t)|H_Y(t-1)) - z_\nu\right] \to 0 \text{ as } \nu \to \infty\,. \tag{8}$$

Consider the functions f_ν given by $f_\nu(u) = \sum_{k=1}^{s_\nu} \sum_{j=1}^{n} \lambda_{k,j}^\nu a_j(u-k)$ for $\nu = 1, 2, \ldots$. Clearly $f_\nu \in A$, and it is easy to show that $h_i \in L_n^2$. From these observations, it is clear that we can write

$$||h_i - f_\nu||^2 = \text{var}\left[\int_0^\infty [h_i(u) - f_\nu(u)]\zeta(t-du)\right] =$$

$$\text{var}\left[\eta(Y_i(t)|H_Y(t-1)) - \sum_{k,j} \lambda_{k,j}^\nu \int_0^\infty a_j(u-k)\zeta(t-du)\right] =$$

$$\text{var}\left[\eta(Y_i(t)|H_Y(t-1)) - \sum_{k,j} \lambda_{k,j}^\nu Y_j(t-k)\right] \to 0 \text{ as } \nu \to \infty$$

where we have used (7), the fact that $\int_0^\infty a_j(u-k)\zeta(t-du) = \int_0^\infty a_j(u)\,\zeta(t-k-du) = Y_j(t-k)$, and (8). Therefore, $f_\nu \to h_i$ and, since A is a closed set, $h_i \in A$. ∎

This proposition tells us that h_i is (very close to) a function of the form $\sum_{k=1}^\infty \lambda_k a(u-k)$, $\lambda_k \in R^n$, with the λ_k's chosen so as to make h_i "as close as possible" to a_i (where "close" is in terms of the distance $\int ||h_i - a_i||^2$).[5]

Once we have characterized h, we can find c in equation (6) by setting $c = a - h$.[6]

The characterization in Proposition 1 tells us the nature of h, and therefore of c, as functions in L_n^2, and allows us to handle examples quite easily. Later in this paper, we state some other properties of h.

The next proposition is very easy to prove, and is stated mainly for future reference.

Proposition 2: For any given $i = 1, \dots n$, the following are equivalent:

$$\begin{aligned} &i) \quad c_i(u) = 0 \quad \text{for almost every} \quad u \geq 1 \\ &ii) \quad \eta(Y_i(t)|H_Y(t-1)) = \eta(y_i(t)|H_y(t-1)) \\ &iii) \quad a_i(u) = \eta(a_i|A)(u) \quad \text{for almost every } u \geq 1 \, . \end{aligned}$$

Proof: For $i) \Rightarrow ii)$, use

$$y_i(t) = \int_0^1 a_i(u)\zeta(t-du) + \eta(y_i(t)|H_y(t-1))$$

$$Y_i(t) = \int_0^\infty c_i(u)\zeta(t-du) + \eta(Y_i(t)|H_Y(t-1)) \, .$$

Now, since $a = c$ in the interval $[0, 1)$ we get $ii)$ by equating the right hand sides to these two equations.

To show $ii) \Rightarrow iii)$, we note that $ii)$ implies $\int_1^\infty h_i d\zeta = \int_1^\infty a_i d\zeta$, and use the comment in footnote (6).

That $iii) \Rightarrow i)$, is obvious, since $c_i = a_i - \eta(a_i|A)$. ∎

3. MAR of the Sampled Process

In the last section, we found a characterization of the one-step-ahead innovation in discrete time in terms of the underlying continuous time process. We are now in a position to characterize the MAR coefficients of Y.

Proposition 3: The matrices A_k in the MAR of Y are given by:

$$A_k = \left[\int_0^\infty a(u+k)c(u)'du\right] \cdot \left[\int_0^\infty c(u)\,c(u)'du\right]^{-1} \tag{9}$$

where $c = a - \eta(a \mid A) \equiv a - h$.

Proof: Let $V = E\left[\epsilon(t) \cdot \epsilon(t)'\right]$. Now,

$$A_k V = E\Big[Y(t+k)\cdot\epsilon(t)'\Big] = E\Big[\int_0^\infty a(u)\zeta(t+k-du)\cdot\int_0^\infty c(u)'\zeta(t-du)\Big] =$$

$$E\Big[\int_0^\infty a(u)\zeta(t+k-du)\cdot\int_0^\infty c(u-k)'\zeta(t+k-du)\Big] = \int_0^\infty a(u)c(u-k)'du =$$

$$\int_0^\infty a(u+k)c(u)'du$$

where the first equality is easily derived by expressing Y in its MAR, and where we use (6). Finally, since

$$V = E\left[\epsilon(t)\cdot\epsilon(t)'\right] = \int_0^\infty c(u)c(u)'du$$

and the assumption that y is full rank guarantees that V is invertible, we have shown the proposition. ∎

In convolution notation, this result can be expressed as:

$$\text{(10)} \qquad A_k = \left[a^{\star}c^{\top}(k)\right]\left[c^{\star}c^{\top}(0)\right]^{-1} \quad \text{for all } k = 0, 1, \ldots$$

where $c(u)^{\top} = c(-u)'$ for all u.

Proposition 3 tells us that A_k is a weighted average of the function a over the interval $[k, \infty)$ with c as the weighting kernel.

We saw in the last section that $c_i = a_i - \eta(a_i|A)$, so that the ith row of c is the error made when we project a_i on the set A. It is easy to show that for any function f in A, $f(u) = [0, \ldots, 0]$ for all $u \in [0, 1)$,[7] so that $\eta(a_i|A)(u) = 0$ and $c_i(u) = a_i(u)$ for $u \in [0, 1)$. For $u \geq 1$, if $\eta(a_i|A)$ is any good in approximating a_i, we would expect $c_i(u)$ to be small, and the graph of the elements of c_i will look more or less like Figure 1:

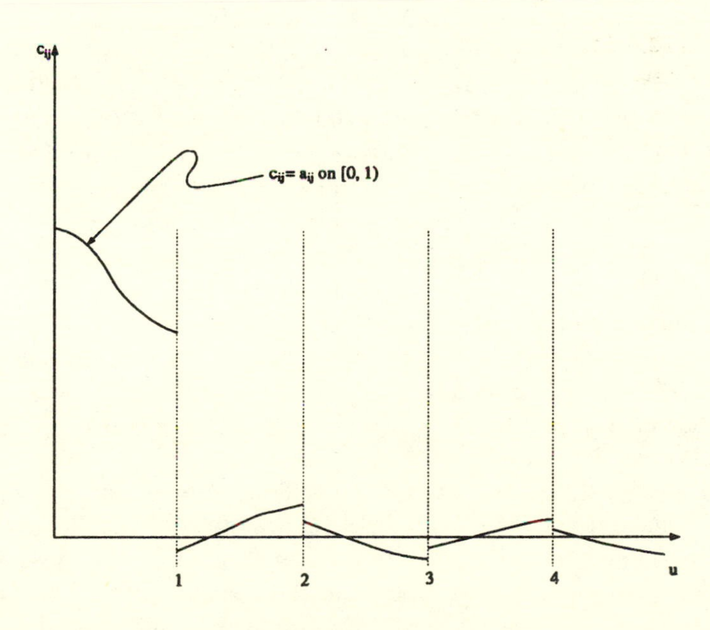

Figure 1.

This illustrates the fact that c will in general give most of the weight to values of a on the interval $[k, k+1)$. On the other hand, c will give more or less weight to values of a_{ij} on the interval $[k+1, \infty)$ depending on how large c_{ij} is in the interval $[1, \infty)$; but since

$$\int_1^\infty ||c_i||^2 = \text{var}\left[\eta(Y_i(t)|H_Y(t-1)) - \eta(y_i(t)|H_y(t-1))\right] ;$$

we have that c_i is small on the interval $[1, \infty)$ when the projection in discrete time approximates well the projection in continuous time. In this case, A_k depends largely on values of a in the interval $[k, k+1)$. In the extreme case that those two projections are equal, Proposition 2 applies, and:

$$A_k = \left[\int_0^1 a(u+k)a(u)'du\right] \cdot \left[\int_0^1 a(u)\,a(u)'du\right]^{-1} \tag{11}$$

so that A_k only depends on values of a on the interval $[k, k+1)$.

There are certain similarities between formula (10) and the formulas that Sims and Geweke obtain in their work on time aggregation of the distributed lag model. In particular, in the distributed lag model the discrete parameters are also obtained by applying a weighting kernel to the continuous time parameters. However, in our case, the kernel in (10) is one-sided, and in general discontinuous, while the opposite was true in the distributed lag model.

Next, we are going to substantiate our claim about the non-continuity of c. For this purpose, it is enough to think of the case of a univariate process. Assume that a is continuous everywhere except at zero (in other words, $a(0) \neq 0$). Since $c(u) = a(u) - \sum_{k=1}^{\infty} \lambda_k a(u-k)$, for any integer, ν, $c(\nu) = a(\nu) - \sum_{k=1}^{\nu} \lambda_k a(\nu - k)$, because $a(\nu - k) = 0$ for $k > \nu$. Therefore, at $u = \nu$ the function c is the sum of $\nu + 1$ functions such that one of them is discontinuous at $u = \nu$ (because, for $k = \nu$ in the summation sign above, $\lambda_\nu a(\nu - \nu) = \lambda_\nu a(0)$, and we assumed that a was discontinuous at $u = 0$) and the remaining ν functions are continuous at $u = \nu$. Therefore, if $\lambda_k > 0$ for all k, c will be discontinuous at all integers.

Equation (10) shows how the coefficients in the ith row of A_k (i.e., the coefficients of the ith variable Y_i) are affected by all the rows in a, so that the ith row of A_k will in general depend on the moving average coefficients in continuous time of all the elements of y. This is the phenomenon that Geweke called "contamination", and that also appeared in the model he studied. The coefficients A_k will be "contaminated" even when the projections in continuous and discrete time coincide, as formula (11) shows. The only general case in which contamination disappears, is when $a_{ij} \equiv 0$ for all $i = j$ (i.e., when the two matrices to the right of (10) are both diagonal). This is a very special case, since it amounts to assuming that $E(y_i(t) \cdot y_j(t')) = 0$ for all $i \neq j$, and all $t, t' \in \mathbf{R}$.

The rest of this section discusses what distortions can be generated by temporal aggregation in the MAR in view of the above results.

We begin by displaying one type of distortion that will be present in most cases. For any integer ν, we can write

$$\operatorname{var}\Big(Y_i(t) - \eta(Y_i(t)|H_Y(t-\nu))\Big) \geq \operatorname{var}\Big(y_i(t) - \eta(y_i(t)|H_y(t-\nu))\Big)$$

and

$$\sum_{k=0}^{\nu-1} A_{k_i} \Sigma A'_{k_i} \geq \int_0^{\nu} a_i(u) a_i(u)' du \ ,$$

where A_{k_i} is the ith row of A_k. In the one variable case, this can be written as

$$\sigma_\epsilon^2 \sum_{k=0}^{\nu-1} (A_k)^2 \geq \int_0^{\nu} a(u)^2 du \ .$$

One way to interpret this is that the first coefficients of the discrete MAR will be too large in absolute value so that, for a univariate model, if we plot the discrete and continuous MAR's in the same graph, we will obtain a version of Figure 2.

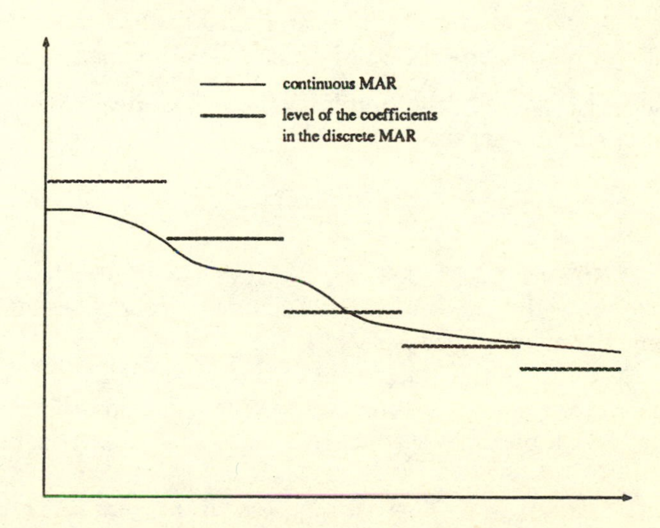

Figure 2.

We could distinguish two reasons why the discrete MAR can be a bad approximation of the continuous MAR; one is contamination, and the other A_k depending on values of a on the interval $[k+1, \infty)$. We now give examples that illustrate the type of distortions the second "reason" can generate, and in what cases this distortion will be severe. We will only consider one-variable examples, in this section, since their effect is present in these cases. These examples, illustrate how Propositions 1, 2 and 3 can be used to analyze time aggregation.

Examples 3.1 and 3.2 display two cases in which the projections in continuous and in discrete time coincide, so that $c(u) = 0$ for $u \in [1, \infty)$, and the shape of A_k is similar to the shape of a. In Example 3.3 we find a very simple MAR in continuous time that will have a distorted discrete MAR. We have said that when c is zero in the interval $[1, \infty)$, A_k is an average of a in the interval $[k, k+1)$; but even in this case, if c is positive *and* negative in the interval $[0, 1)$, A_k will not be a proper average of a in the interval $[0, 1)$; we illustrate this point in Example 3.4. Example 3.5 effectively shows one way to construct continuous MAR's that will generate distorted discrete MAR's. Example 3.6 discusses the effects of unit-averaging. Even though the main message of Example 3.6 is that when $a(0) = 0$ and a is continuous, we may expect distortions in the discrete MAR, Example 3.7 shows that this is not always true. Finally, Examples 3.1, 3.4 and 3.7 are concrete illustrations of the aliasing problem. They show three different continuous MAR's that generate exactly the same sampled process: in particular, Y is an autoregressive process of order one in all these examples.

<u>Example 3.1:</u> Probably the simplest case we can deal with is the AR(1) process in continuous time

$$y(t) = \int_0^\infty e^{-\lambda u}\zeta(t - du) \qquad t \in \mathbf{R}, \quad \lambda > 0 \ .$$

In this case, $e^{-\lambda}y(t-1) = \int_1^\infty e^{-\lambda u}\zeta(t-du) = \eta(y(t)|H_y(t-1))$, and $e^{-\lambda u}\, y(t-1) \in H_Y(t-1)$; the orthogonality conditions in continuous time imply that, in particular, $[Y(t) - e^{-\lambda}Y(t-1)] \perp Y(t-k)$ for all $k = 1, 2, \ldots$, so that $e^{-\lambda}Y(t-1)$ *is* the projection in discrete time.

Therefore, Y has an autoregressive representation (henceforth ARR) of order 1, with parameter equal to $e^{-\lambda}$, so that the MAR of Y is given by:

$$A_k = e^{-\lambda k} = a(k) \qquad \text{for all integers } k \ .$$

For illustrative purposes, we next find A_k by using propositions 1, 2, 3. Since $e^{-\lambda}a(u-1) = a(u)$ for all $u \geq 1$, clearly $e^{-\lambda}a(u-1) =$

$\eta(a|A)(u)$ for all $u \geq 1$, so that Proposition 2 applies, and we can use formula (11) to conclude:

$$A_k = \left[\int_0^1 e^{-\lambda(u+k)}\, e^{-\lambda u}\, du\right] \cdot \left[\int_0^1 e^{-\lambda 2u}\, du\right]^{-1} = e^{-\lambda k}$$

Example 3.2: If a is constant over intervals $[k,\ k+1)$ and y one-dimensional, letting $\epsilon(t) = \int_0^1 \zeta(t-du)$ and $A_k = a(k)$, we can write:

$$Y(t) = \sum_{k=0}^{\infty} \int_k^{k+1} a(u)\zeta(t-du) = \sum_{k=0}^{\infty} a(k) \int_k^{k+1} \zeta(t-du) = \sum_{k=0}^{\infty} A_k \epsilon(t-k)\,.$$

This expresses Y in moving average form. It has to be true that the Fourier transform of a has no zeroes on the right half of the complex plane. This guarantees that the Fourier transform of $\{A_k\}$ has no zeroes inside the unit disk, and that this is associated with the Wold representation of Y.

In this case, as in the previous example, the discrete MAR equals the continuous MAR sampled at integers, and from the way ϵ was chosen, it is clear that the one-step-ahead innovations in discrete and continuous time coincide, so that the one-step-ahead projections in discrete and continuous time are equal.

Example 3.3: Hansen and Sargent (1984) showed that for a MAR a that satisfies:

— a is continuous at all $u \geq 0$
— $a(0) \neq 0$
— Y has an ARR,

if $c = 0$ almost everywhere in $[1,\ \infty)$ then Y has an ARR of order 1.

In our framework, this is displayed in the following way: if Y had an autoregressive representation of order more than 1, some λ_k would be different from zero for $k > 1$. Then $\eta(a|A)$ would be discontinuous at $u = k$ (this can be deduced by using the line of argument on page 243 showing that c is in general discontinuous), so it is not possible that $\eta(a|A)(u) = a(u)$ for a.e. $u \in [1,\ \infty)$. Thus, $c \neq 0$ in a subset of $[1,\ \infty)$ of positive measure.

This implies that apparently well behaved a's will yield distorted A_k's. For example, the MAR given by : $a(u) = e^{-\lambda_1 u} + e^{-\lambda_2 u}$ has a function c with $\int_1^{\infty} ||c||^2 > 0$. Indeed, if this was not true, then $c \equiv 0$ almost everywhere in $[1,\ \infty)$, and by Proposition 2 there would exist a constant μ such that $\mu\, a\,(u-1) = a(u)$ for all $u \geq 1$. In particular, for $u = 1$ and $u = 2$ this would imply

$$\mu = e^{-\lambda_1} + e^{-\lambda_2} \qquad \text{and} \qquad \mu(e^{-\lambda_1} + e^{-\lambda_2}) = e^{-\lambda_1 2} + e^{-\lambda_2 2},$$

which in turn would imply $(e^{-\lambda_1} + e^{-\lambda_2})^2 = e^{-\lambda_1 2} + e^{-\lambda_2 2}$, which is not possible.

Example 3.4: Let a be as depicted in Figure 3.

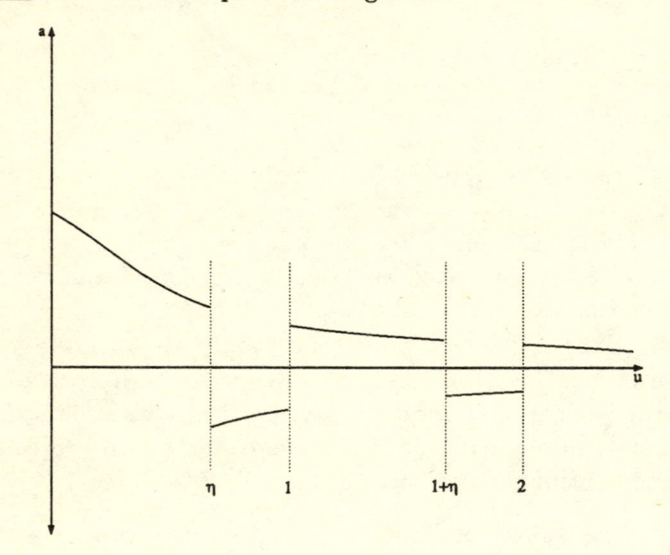

Figure 3.

In Figure 3, $|a(u)| = e^{-\lambda u}$ for all $u \geq 0$, but for each integer k, if $u \in [k+\eta,\, k+1)$, then $a(u) < 0$.

Again, $e^{-\lambda} a(u-1) = a(u)$ for all $u \geq 1$, so that, by Proposition 2, $\eta(Y_i(t)|H_Y(t-1)) = e^{-\lambda} Y(t)$. Note that $A_k = e^{-\lambda k}$, and the discrete MAR fails completely to capture the oscillations in a.

The interest of this example is to show that it is possible for the projection with discrete data to be as good as the projection in continuous time, and yet the discrete time MAR be very different from the continuous MAR. Indeed, in this case we can apply Proposition 2 to conclude that the two projections coincide, but the discrete MAR fails completely to capture the oscillations in the continuous MAR. We may expect this to happen when a is both positive and negative in large parts of the interval $[0, 1)$. Recall that, from formula (11), we know that A_k is largely a weighted average of a on the interval $[k, k+1)$, where the weights are given precisely by a on the interval $[0, 1)$; therefore, if a takes both positive and negative values, A_k is not a proper average of a on the interval $[k, k+1)$.

Example 3.5: For $\{A_k\}$ not to be distorted, it is necessary that:

$$\int_1^2 c^2 = \int_1^2 (a(u) - h(u))^2 du = \int_1^2 (a(u) - \mu \cdot a(u-1))^2 du = 0$$

so that, unless there exits some constant μ such that $a(u) = \mu\, a(u-1)$ for a.e. $u \in [1, 2)$, we will have distortions of the discrete MAR. Keeping this in mind, it is quite easy to generate examples in which $\int_1^2 c^2$ is very large, by considering functions a that look very different in the interval $[0, 1)$ and in the interval $[1, 2)$. The graphs in Figure 4 illustrate this situation.

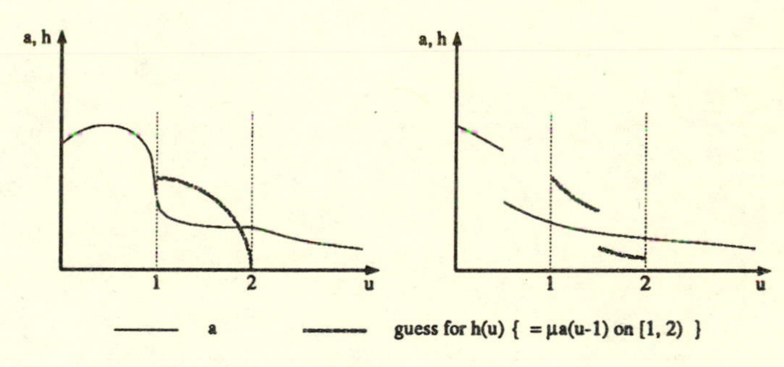

Figure 4.

Example 3.6: It has been suspected for a long time that when a is continuous as a function defined on the whole real line (i.e., when $a(0) = 0$), the discrete MAR does not approximate a well. Hansen and Sargent (1984), have an example with large distortions. We will now argue that this is in fact the case if a is continuous and positive at $u = 1$.

We have argued before that for $u \in [1, 2)$ $h(u) = \mu\ a(u-1)$ for some coefficient μ. Therefore, if a is continuous and strictly positive for all $u > 0$, but $a(0) = 0$, the graph of a and h will be as depicted in Figure 5.

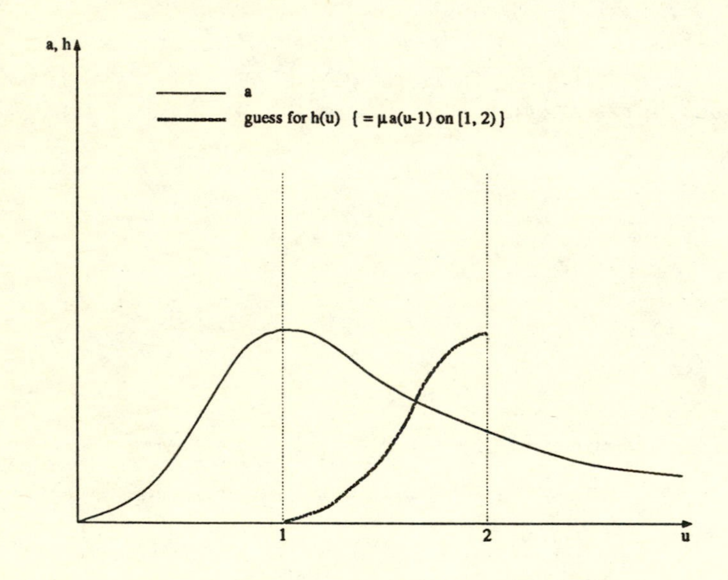

Figure 5.

As illustrated in Figure 5, $\int_1^2 c^2 = \int_1^2 (a-h)^2$ is large in relation to $\int_0^1 c^2 = \int_0^1 a^2$. In this case, A_k will depend more on values of a in the interval $[k+1,\ k+2)$ than on values of a in $[k,\ k+1)$.

Example 3.7: In view of the previous example, we should not conclude that any MAR in continuous time with $a(0) = 0$ will give a bad approximation in discrete time. For example, take a function a that is continuous in the interval [0, 1], satisfies $a(0) = a(1) = 0$, and can be extended to $[1, \infty)$ by setting: $a(u) = \ell a(u-1)$ for some $0 < \ell < 1$, for all $u \in [1, \infty)$; the graph of this function will be as depicted in Figure 6.

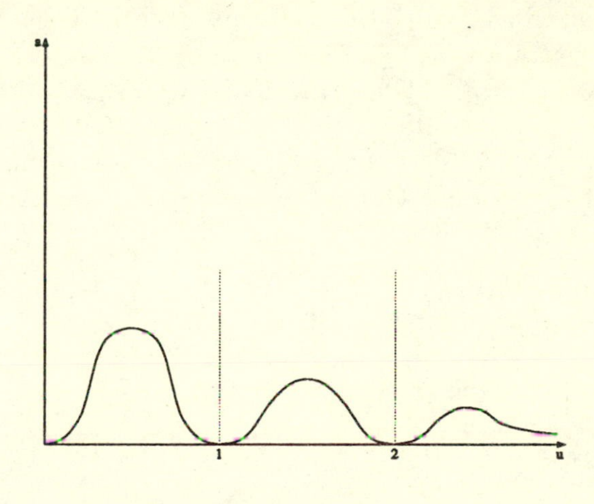

Figure 6.

Because of the way a was defined, clearly Proposition 2 (ii) applies, and

$$\eta(Y(t)|H_Y(t-1)) = \eta(y(t) \mid H_y(t-1)) ,$$

so that Y has an autoregressive representation of order 1.

4. Granger Causality

Consider a bivariate process in continuous time y, such that y_2 does not Granger cause y_1. It is well known that, in general, Y_2 will Granger cause Y_1. It is possible to deduce this by combining results in Sims (1971, 1972b).[8]

There are cases, however, in which the absence of Granger causality from the second variable to the first, *does* carry over to the sampled process. One case in which this happens is when y_1 and y_2 are uncorrelated at all dates; then Y_1 and Y_2 will also be uncorrelated at all dates, so that past Y_2's will not help in predicting Y_1. Another case is given by the following proposition, which says that if the first variable can be predicted with equal accuracy whether we use continuous or discrete data and if y_2 does not Granger cause y_1, then Y_2 does not Granger cause Y_1.

Proposition 4: If $\eta(y_1(t) \mid H_y(t-1)) = \eta(Y_1(t) \mid H_Y(t-1))$ and y_2 fails to Granger cause y_1, then Y_2 will also fail to Granger cause Y_1.

Proof: The conditions in this proposition imply:

$$\begin{aligned}\eta(y_1(t) \mid H_y(t-1)) &= \eta(y_1(t) \mid H_{y_1}(t-1)) \\ &= \eta(Y_1(t) \mid H_Y(t-1)),\end{aligned}$$

therefore $\eta(Y_1(t) \mid H_y(t-1)) \in H_{y1}(t-1) \cap H_Y(t-1)$. By definition, $H_{y_1}(t-1) \cap H_Y(t-1) = H_{Y_1}(t-1)$, so that $\eta(Y_1(t) \mid H_y(t-1)) \in H_{Y_1}(t-1)$ and, since $H_{Y_1}(t-1) \subset H_y(t-1)$, we have that $\eta(Y_1(t) \mid H_y(t-1)) = \eta(Y_1(t) \mid H_{Y_1}(t-1))$. Combining this last equality with the first condition of the proposition tells us that $\eta(Y_1(t) \mid H_{Y_1}(t-1)) = \eta(Y_1(t) \mid H_Y(t-1))$. ∎

This type of result is not surprising. The reason that, in general, Y_2 will Granger cause Y_1 even when past y_2 do not help in predicting y_1, is that the values of y_1 between integers enter in $\eta(Y_1(t) \mid H_y(t-1))$. Then, in the discrete projection, past values of Y_2 could help predict Y_1 because they help predict values of y_1 between integers. However, in the case that the above proposition considers, there is no point in trying to predict the values of y_1 between integers, so that there is no room for past Y_2 to help predict Y_1.

We can construct cases in which the effect of temporal aggregation on Granger causality is the opposite of the effect described at the beginning of this section, that is, Y_2 does not Granger cause Y_1 even though y_2 Granger causes y_1. Consider the MAR in continuous time given by the functions a_{ij} depicted in Figure 7.

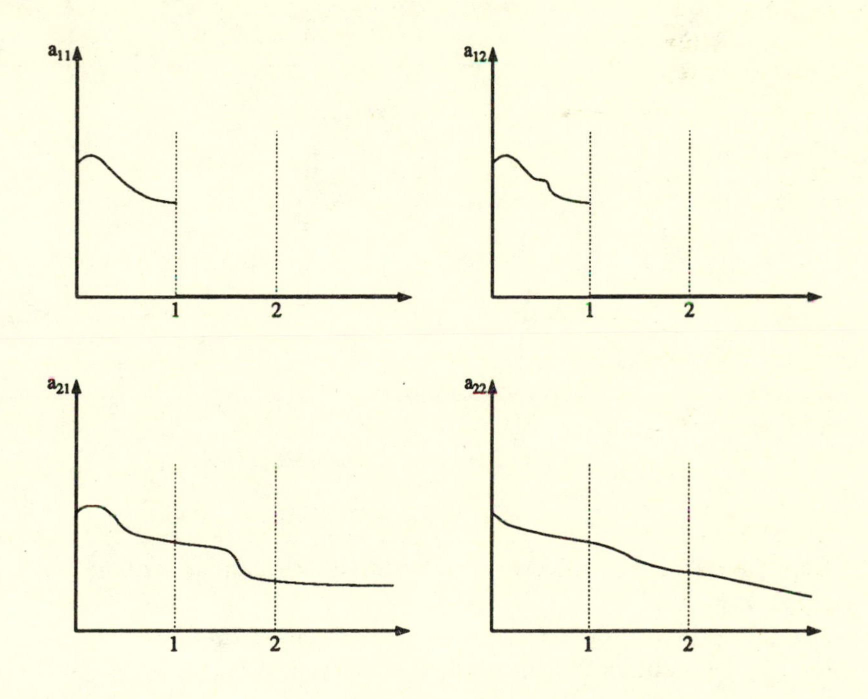

Figure 7.

In this case, clearly $\mathrm{cov}(Y_1(t), Y_2(t-2)) = \int_0^\infty [a_{11}(u)a_{21}(u-s) + a_{12}(u)a_{22}(u-s)]\, du = 0$ for integers $s \geq 1$, and past Y_2 will not help in predicting current Y_1. But, since neither a_{11} nor a_{12} is identically zero, y_2 Granger causes y_1.

Another case in which this happens is shown in Figure 8.

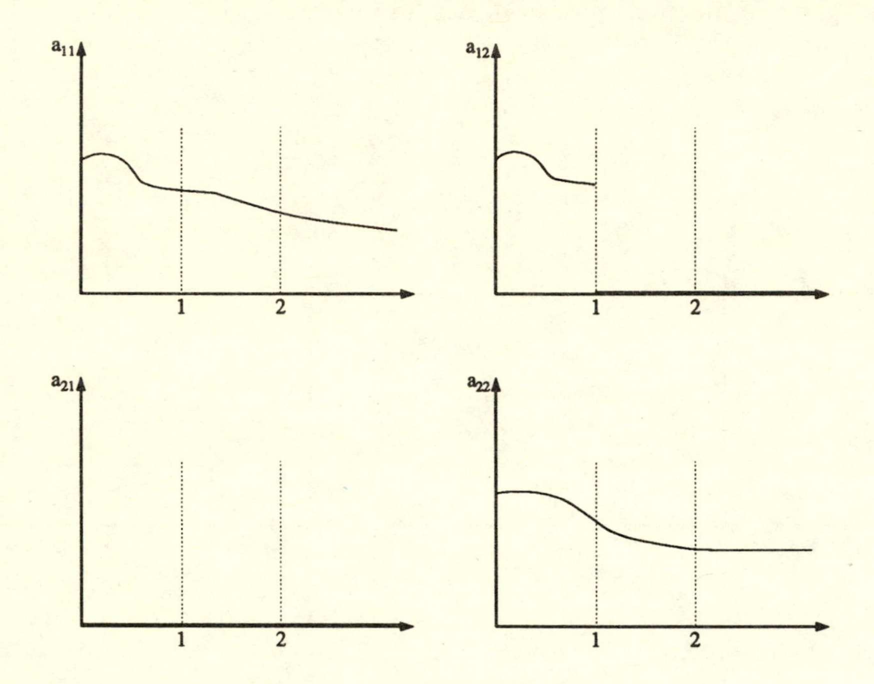

Figure 8.

Again, $Y_1(t)$ will be uncorrelated with past Y_2's and Y_2 will not Granger cause Y_1.

These two examples are very special cases; the characteristic they have in common is that much of the variance of $Y_1(t)$ is due to innovations in the interval $[t, t-1)$.

5. Unit-Averaged Data

Sometimes the data available to us on a given variable consist of averages of observations over a certain period of time. For example, our monthly data on a given series may consist of the average of the weekly data for that month. Say that we have unit-averaged observations on the first m-variables. We can model this situation as follows. We observe the process Y in discrete time given by

$$Y_i(t,\omega) = \int_0^1 y_i(t-s,\omega)ds \qquad i = 1,\ldots,m\ ;$$
$$Y_i(t,\omega) = y_i(t,\omega) \qquad i = m+1,\ldots,n, \quad \text{for all } t \in \mathbf{I}, \quad \text{and all } \omega \in \Omega$$

where the above integral sign refers to Lebesgue integral.

Clearly, any of the unit-averaged variables Y_i , $i \leq m$, can be obtained by sampling the continuous time process $\tilde{y}$ given by:

$$\tilde{y}_i(t) = \int_0^1 y_i(t-s)ds \qquad \text{all } t \in \mathbf{R}\,.$$

In order to apply the results of Section 3, we have to find the MAR of the continuous time process $\tilde{y}$. Let us define the function $\ell : \mathbf{R} \to \mathbf{R}$ as

$$\ell(u) = 1 \quad \text{if } u \in [0,1)$$
$$\ell(u) = 0 \quad \text{otherwise}\,;$$

so that we can write

$$\tilde{y}_i(t) = \int_{-\infty}^{\infty} \ell(u)\, y_i(t-u)\, du$$

and we have that

$$(15)\quad f_{\tilde{y}_i}(\omega) = \hat{\ell}(\omega) f_{y_i}(\omega) \hat{\ell}'(\omega) = \hat{\ell}(\omega)\hat{a}_i(\omega)\hat{a}_i'(\omega)\hat{\ell}'(\omega), \qquad \omega \in [-\pi,\pi]$$

where f_χ denotes the spectral density of χ, and $\hat{b}$ denotes the Fourier transform of the function b; in the above equation we have used the formula for the spectral representation of linear combinations of random variables (see Rozanov (1967)). Since $\hat{\ell}$ has no zeroes in the left half plane, (15) indicates that $\ell^* a_i$ gives the MAR for $\tilde{y}_i$, and we can write

$$\tilde{y}_i(t) = \int_0^{\infty} \tilde{a}_i(u)\zeta(t-du)$$

where

$$(15') \qquad \tilde{a}_i(u) = \ell * a(u) = \int_0^1 a_i(u-s)ds\,,\ t \in \mathbf{R},\ i = 1, \ldots, m\,.$$

The next proposition states that $\tilde{a}$ is always smoother than a. As usual, we denote the space of functions that can be differentiated s times by C^s.

Proposition 5: For any $i,\ j = 1,\ \ldots,\ n$

i) If $a_{ij} \in L^1 \left[\text{i.e., } \int |a_{ij}| < \infty\right]$ then $\tilde{a}_{ij}$ is continuous.

ii) If, in addition, $a_{ij} \in C^s$ then $\tilde{a}_{ij} \in C^{s+1}$.

Proof:

i) Define $F(x) = \int_{-\infty}^{x} a_{ij}(s)ds$; then F is an indefinite integral; therefore, F is continuous.[9] Since $\tilde{a}_{ij}(u) = \int_{u-1}^{u} a_{ij}(s)ds = F(u) - F(u-1)$, $\tilde{a}_{ij}$ is continuous.

ii) If a_{ij} is continuous at x, then F has a derivative at x, and $F'(x) = a_{ij}(x)$.[10] ∎

Note that Proposition 5 refers to $\tilde{a}$ as a function defined on the whole real line, so that, since $\tilde{a}(u) = 0$ for $u < 0$, i) in Proposition 5 implies that $\tilde{a}(0) = 0$. Remember that Example 3.6 dealt with the case of a continuous function a. The comments made there apply to this section.

The rest of this section analyzes the effects of using unit-averaged and sampled data at the same time.

Consider a 2-dimensional process in continuous time; we have sampled observations on the second variable, but we have unit-averaged observations on the first. It is common practice in this case to estimate a 2-variable system in discrete time, consisting of point-in-time observations for one variable, and unit-averaged observations for the other. We are going to argue that this practice will systematically overstate the importance of the second variable (consisting of sampled observations) in predicting the first one.

We will discuss this by giving an informal, general argument, and by displaying several simulations.

The general argument goes as follows. Consider a two-dimensional process y with MAR represented by the graph in Figure 9.

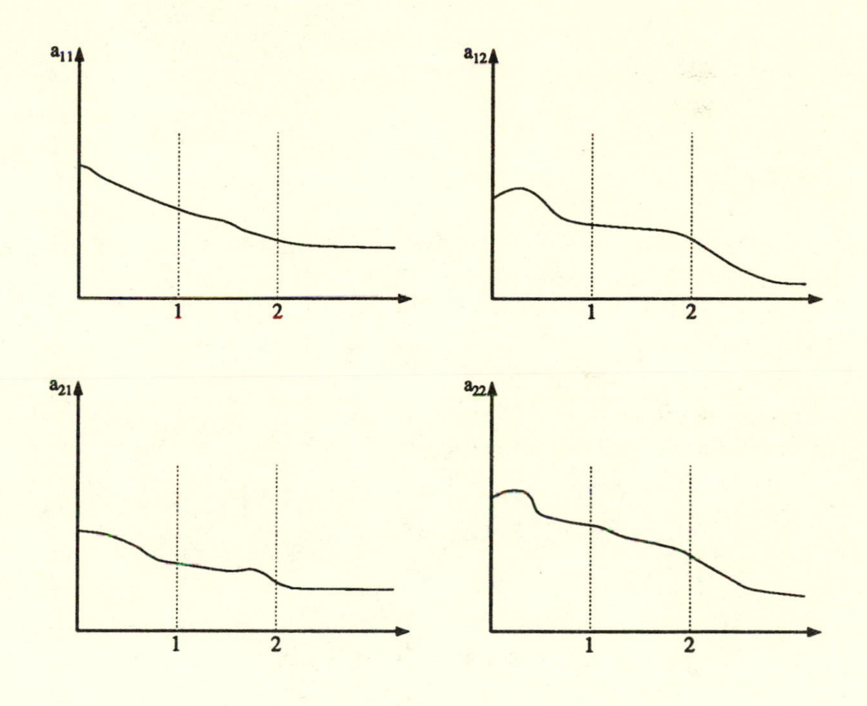

Figure 9.

The discrete system obtained by mixing unit-averaged and sampled data in the way described above, is equivalent to sampling the continuous time process $\hat{y}(t)' = (\tilde{y}_1(t),\ y_2(t))'$. The MAR of $\hat{y}$ is given by $\hat{a}(u) = \begin{bmatrix} \tilde{a}_1(u) \\ a_2(u) \end{bmatrix}$, where $\tilde{a}_1$ is given by equation (15′), so that the graph of $\hat{a}$ will be as in Figure 10.

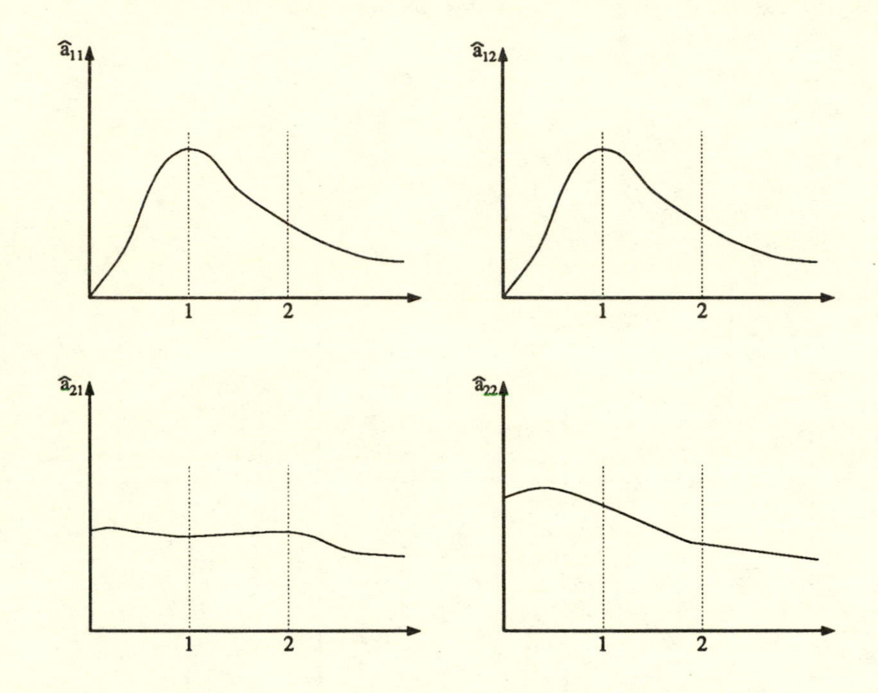

Figure 10.

From the discussion in Section 2 (and the appendix) we have learned that, letting $\mu_k \in \mathbf{R}^2$ be the vectors such that

$$\eta(Y^1(t) \mid H_Y(t-1)) = \sum_{k=1}^{\infty} \mu_k Y(t-k) ,$$

the μ's are those vectors that make the function $h_1(u) = \sum_{k=1}^{\infty} \mu_k\, a(u-k)$ as close as possible to the function a_1.

We will argue that, if Y_1 is unit-averaged, the portion of the variance of Y_1 explained by Y_2 will be larger than if both variables were sampled, due to the fact that the first elements of the vectors μ_k will, in general, be smaller.

We will look, first, at the sampled system. Let us consider how much the first elements of μ_k can contribute to make h_1 close to a_1. Assume we set the second element of μ_k equal to zero, and we try to approximate a_1 with a function of the type $g(u) = \sum \lambda_k a_1(u-k)$.

Then, g has to look like Figure 11.

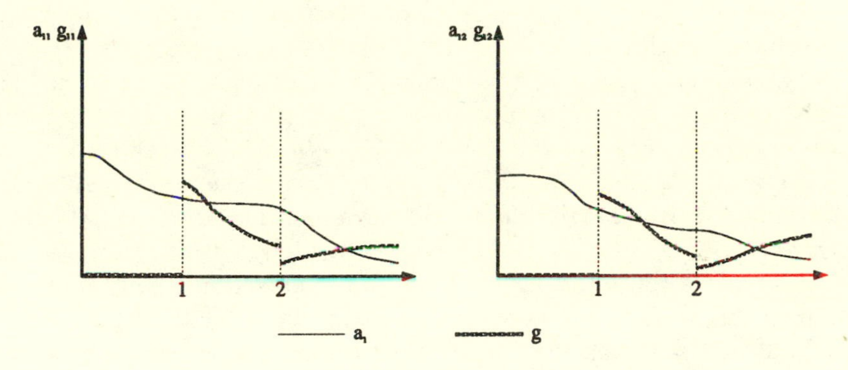

Figure 11.

However, if we did the same with $\hat{a}$ instead of a, and let $\hat{g}(u) = \sum_{k=1}^{\infty} \lambda_k \hat{a}_1(u-k)$ and then compare $\hat{g}$ with $\hat{a}_1$, we have a version of Figure 12 (see Example 3.6).

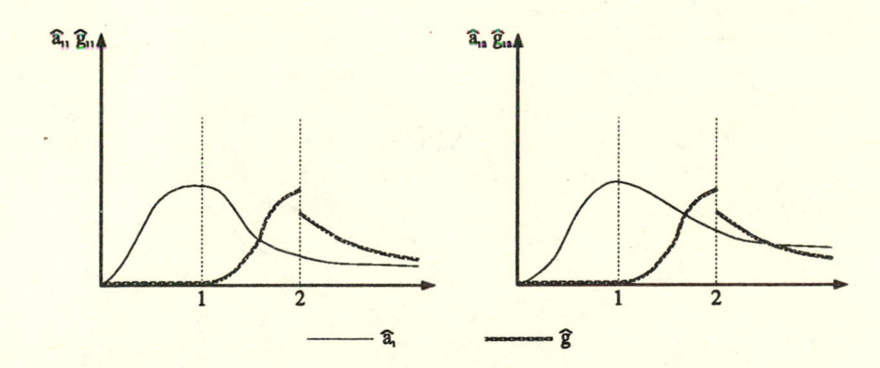

Figure 12.

The point of these graphs is to demonstrate that we can approximate a_1 better with a function like g than we can approximate $\hat{a}_1$ with a function like $\hat{g}$.

Now, since the vectors μ_k are chosen in order to make h_1 as close to a as possible, we would expect that after unit-averaging the first elements of μ_k are smaller. Therefore, past Y_1 will not be as helpful in predicting $Y_1(t)$ if Y_1 is unit-averaged.[11]

The same intuition can be used to justify the claim that if only Y_1 is unit-averaged, past Y_1 will not be as effective in predicting Y_2 as in the case that both variables are sampled, and that the same is true if we compare the mixed system with the system in which both variables are averaged.

Next, we will present a few simulations that support our claims. We have simulated several AR(2), discrete time processes with two variables. Then we have temporarily aggregated them in three ways: the "sampled system" is obtained by sampling both variables every three periods; the "averaged system" is obtained by taking the average of each variable over three periods; finally, the "mixed system" is obtained by averaging the first and sampling the second variable every three periods. This does not correspond exactly with the situation discussed above, since the variables are not averaged in a continuous way, but it can be interpreted as an approximation; furthermore, it corresponds to a situation that arises often in practice, when the researcher has two variables recorded at different time-intervals (say monthly and quarterly) and has to aggregate over time one of the variables.

We have run a VAR for each of these temporally aggregated processes. In the tables below we report the decomposition of variance for each process. Since the variance decomposition can be affected by the order in which the Choleski decomposition is obtained, we have computed the decomposition for both orderings in all the cases.[12]

If our claims are true, the percentage variance of Y_1 and Y_2 explained by Y_1 would be smaller in the "mixed" than in the other two systems. In these simulations, the effect of mixing averaged and sampled data is almost always the one predicted on the last page, and in some cases the changes are very large (particularly in Table 1).

Only in one out of the six simulations we report did the change in the decomposition of variance of one of the variables go clearly in the other direction (Table 2, first ordering, decomposition of Y_1). Far from proving a theorem, we have just presented a heuristic argument, so that, probably, our intuition is not true for all processes. On the other hand, the result in Table 2 may be due to sampling error or to the fact that, as we pointed before, the type of unit-averaging that we have performed in the simulations is not quite the continuous time averaging discussed in the rest of this section.

In any event, these simulations do support the claim that, in general, by unit-averaging the first variable, Y_1 becomes less important in determining both Y_1 and Y_2.

Table 1

$$\begin{bmatrix} y_1(t) \\ y_2(t) \end{bmatrix} = y(t) = \begin{bmatrix} 1.04, & .65 \\ .59, & .98 \end{bmatrix} y(t-1) + \begin{bmatrix} -.36, & -.33 \\ -.29, & -.33 \end{bmatrix} y(t-2) + \varepsilon(t)$$

Ordering 1: Y_1, Y_2

Variance of ...	Percentage explained by Y_1 in each system		
	Y_1, Y_2 sampled	Y_1, Y_2 averaged	Y_1 aver. Y_2 sampled
Y_1	56	35	15
Y_2	23	10	4

Ordering 2: Y_2, Y_1

Variance of ...	Percentage explained by Y_1 in each system		
	Y_1, Y_2 sampled	Y_1, Y_2 averaged	Y_1 aver. Y_2 sampled
Y_1	85	93	.5
Y_2	85	92	.2

Table 2

$$\begin{bmatrix} y_1(t) \\ y_2(t) \end{bmatrix} = y(t) = \begin{bmatrix} 1.37, & -.03 \\ .41, & 1.06 \end{bmatrix} y(t-1) + \begin{bmatrix} -.44, & .02 \\ -.28, & -.23 \end{bmatrix} y(t-2) + \varepsilon(t)$$

Ordering 1: Y_1, Y_2

Variance of ...	Percentage explained by Y_1 in each system		
	Y_1, Y_2 sampled	Y_1, Y_2 averaged	Y_1 aver. Y_2 sampled
Y_1	87	92	95
Y_2	49	55	40

Ordering 2: Y_2, Y_1

Variance of ...	Percentage explained by Y_1 in each system		
	Y_1, Y_2 sampled	Y_1, Y_2 averaged	Y_1 aver. Y_2 sampled
Y_1	97	97	97
Y_2	37	40	21

Table 3

$$\begin{bmatrix} y_1(t) \\ y_2(t) \end{bmatrix} = y(t) = \begin{bmatrix} 1.2\,, & .5 \\ .2\,, & .9 \end{bmatrix} y(t-1) + \begin{bmatrix} -.4\,, & -.2 \\ -.1\,, & -.2 \end{bmatrix} y(t-2) + \varepsilon(t)$$

Ordering 1: Y_1, Y_2

Variance of ...	**Percentage explained by Y_1 in each system**		
	Y_1, Y_2 sampled	Y_1, Y_2 averaged	Y_1 aver. Y_2 sampled
Y_1	73	66	48
Y_2	8	2	2

Ordering 2: Y_2, Y_1

Variance of ...	**Percentage explained by Y_1 in each system**		
	Y_1, Y_2 sampled	Y_1, Y_2 averaged	Y_1 aver. Y_2 sampled
Y_1	55	57	36
Y_2	10	10	4

Table 4

$$\begin{bmatrix} y_1(t) \\ y_2(t) \end{bmatrix} = y(t) = \begin{bmatrix} 1.09, & .33 \\ .29, & 1.06 \end{bmatrix} y(t-1) + \begin{bmatrix} -.32, & -.18 \\ -.16, & -.3 \end{bmatrix} y(t-2) + \varepsilon(t)$$

Ordering 1: Y_1, Y_2

Variance of ...	Percentage explained by Y_1 in each system		
	Y_1, Y_2 sampled	Y_1, Y_2 averaged	Y_1 aver. Y_2 sampled
Y_1	81	87	67
Y_2	21	26	9

Ordering 2: Y_2, Y_1

Variance of ...	Percentage explained by Y_1 in each system		
	Y_1, Y_2 sampled	Y_1, Y_2 averaged	Y_1 aver. Y_2 sampled
Y_1	80	83	61
Y_2	23	21	8

Table 5

$$\begin{bmatrix} y_1(t) \\ y_2(t) \end{bmatrix} = y(t) = \begin{bmatrix} .43, & .21 \\ .34, & .29 \end{bmatrix} y(t-1) + \begin{bmatrix} .14, & .03 \\ .05, & .12 \end{bmatrix} y(t-2) + \varepsilon(t)$$

Ordering 1: Y_1, Y_2

Variance of ...	Percentage explained by Y_1 in each system		
	Y_1, Y_2 sampled	Y_1, Y_2 averaged	Y_1 aver. Y_2 sampled
Y_1	89	84	87
Y_2	20	23	23

Ordering 2: Y_2, Y_1

Variance of ...	Percentage explained by Y_1 in each system		
	Y_1, Y_2 sampled	Y_1, Y_2 averaged	Y_1 aver. Y_2 sampled
Y_1	95	88	86
Y_2	15	20	17

Table 6

$$\begin{bmatrix} y_1(t) \\ y_2(t) \end{bmatrix} = y(t) = \begin{bmatrix} .37, & .26 \\ .28, & .34 \end{bmatrix} y(t-1) + \begin{bmatrix} .13, & .04 \\ .04, & .12 \end{bmatrix} y(t-2) + \varepsilon(t)$$

Ordering 1: Y_1, Y_2

Variance of ...	**Percentage explained by Y_1 in each system**		
	Y_1, Y_2 sampled	Y_1, Y_2 averaged	Y_1 aver. Y_2 sampled
Y_1	90	93	78
Y_2	20	36	15

Ordering 2: Y_2, Y_1

Variance of ...	**Percentage explained by Y_1 in each system**		
	Y_1, Y_2 sampled	Y_1, Y_2 averaged	Y_1 aver. Y_2 sampled
Y_1	88	86	81
Y_2	18	31	18

6. Convergence of the Discrete MAR as the Sampling Interval Goes to Zero

For each $\delta > 0$, we define a discrete stochastic process Y^δ on our original probability space, by setting

$$Y^\delta(t,\omega) = y(t\delta,\omega) \quad \text{for all } t \in \mathbf{I}, \text{ all } \omega \in \Omega$$

and let the MAR of Y^δ be:

$$Y^\delta(t) = \sum_{k=0}^{\infty} A_k^\delta \epsilon^\delta(t-k)\,. \tag{17}$$

We are interested in determining under what conditions and in what sense the δ-discrete MAR converges to the continuous MAR, and the discrete innovation converges to the continuous one, as the sampling interval δ goes to zero.

Note that the way we have defined $Y^\delta(t)$, this 't' does not correspond to the point 't' in real time (instead, it corresponds to the date '$t\delta$').

We will change our notation slightly in this section, by setting:

$$H_{Y^\delta}(q) = \text{cl}\left\{ \text{linear subspace spanned by } Y^\delta(t) \text{ for } \delta t \le q \right\} \text{ for } q \in R$$

so that 'q' now refers to the corresponding date *in real time*. Actually, $Y^\delta(q)$ itself may not be defined (when q/δ is not an integer). We will write $Y^\delta(t/\delta)$ regularly, assuming implicitly that this is well defined, that is, assuming that t/δ is an integer.

One type of convergence that will prove useful for our purposes and that has interest in itself, is whether, for a given $\alpha > 0$, the projection of $y(t+\alpha)$ on information up to t in continuous time, can be approximated arbitrarily well by observations of Y^δ up to 't' ('t' a date in real time), as δ goes to zero. The next Proposition deals with this question.

In this section we will use repeatedly the following property of Hilbert spaces: let S be such a space; for any sequence $\{x_\nu\}$ in S such that x_ν converges to x (in the distance induced by the inner product on S) then $\langle x_\nu, y\rangle \to \langle x, y\rangle$ as $\nu \to \infty$ for all $y \in S$.

Let $B : \mathbf{R} \to \mathbf{R}^{n\times n}$ be the autocovariance function of y, defined by:

$$B(s) = E(y(t+s)\cdot y(t)'), \quad s \in \mathbf{R}\,.$$

<u>Lemma 3:</u> Let y satisfy assumptions 1, 2, 3 in the introduction, then $y(t+s) \to y(t)$ in mean square as $s \to 0$. This in turn implies that B is continuous at 0.

Proof: For any $i = 1, \ldots, n$, using the MAR of y, we have

$$E(y_i(t) - y_i(t+s))^2 = \int_0^\infty ||a_i(u) - a_i(u+s)||^2 du =$$
$$= \sum_{j=1}^{n} \int_0^\infty |a_{ij}(u) - a_{ij}(u+s)|^2 \, du \; .$$

By Theorem 8.19 in Wheeden and Zygmund (1977), we know that any square integrable function $f : R \to R$, has the property that $\int |f(u) - f(u+s)|^2$ goes to zero as s goes to zero. Since each a_{ij} is square integrable, each element in the above sum goes to zero as $s \to 0$. Thus $y_i(t+s) \to y_i(t)$ in mean square.

Therefore, by the property of Hilbert spaces stated above, we have:

$$B_{ij}(s) = E(y_i(t) \cdot y_j(t+s)') \to E(y_i(t) \cdot y_j(t)') = B_{ij}(0) \text{ as } s \to 0$$

and B is continuous at 0. ▮

Proposition 6: If y satisfies assumptions 1, 2, 3 in the introduction, then

$$\eta\left[Y_i^\delta(t/\delta) \mid H_{Y^\delta}(t-\alpha)\right] \to \eta\left[y_i(t) \mid H_y(t-\alpha)\right] \quad \text{as } \delta \downarrow 0 \; .$$

for any $i = 1, \ldots, n$, $\alpha > 0$, and $t \in \mathbf{R}$.

Proof: Because of stationarity, it is enough to show the theorem for $t = 0$.

By definition of $H_y(-\alpha)$, there exists a sequence of random variables $\{a_\nu\}$ such that each a_ν can be written as

$$a_\nu = \sum_{k=1}^{s_\nu} \left[\lambda_k^\nu\right]' \cdot y(t_k^\nu) \quad \text{for some } \lambda_k^\nu \in \mathbf{R}^n,\ s_\nu \text{ finite, and } t_k^\nu < -\alpha \; ,$$

with the property that $a_\nu \to \eta(y_i(0)|H_y(-\alpha))$ as $\nu \to \infty$.

Now choose any $\theta > 0$, and choose ν' such that

$$\left[E(a_\nu - \eta(y_i(0) \mid H_y(-\alpha)))^2\right]^{\frac{1}{2}} < \theta/2 \; .$$

First, we want to show that for each δ, we can select an element b_δ of $H_{Y^\delta}(-\alpha)$ such that $b_\delta \to a_\nu$, as $\delta \downarrow 0$ (remember that we already fixed ν'). Set:

$$b_\delta = \sum_{k=1}^{s_{\nu'}} \left[\lambda_k^{\nu'}\right]' y(t_k^\delta)$$

where t_k^δ is the closest number to $t_k^{\nu'}$ that is divisible by δ; here $t_k^{\nu'}$, $\lambda_k^{\nu'}$ and s_ν are those in the above expression for a_ν.

Using the previous lemma, since $t_k^\delta \to t_k^{\nu'}$, clearly $y(t_k^\delta) \to (t_k^{\nu'})$ in mean square as $\delta \downarrow 0$, so that $b_\delta \to a_{\nu'}$. Hence, for the θ chosen above, there exists some δ' such that if $\delta < \delta'$ we can apply the triangle inequality to conclude that

$$\left[E\left(b_\delta - \eta(y_i(0)|H_y(-\alpha))^2\right]^{\frac{1}{2}} \equiv d\left[b_\delta, \eta(y_i(0)|H_y(-\alpha))\right] \leq$$

$$\leq d\left[b_\delta, a_{\nu},\right] + d\left[a_\nu, \eta(y_i(0)|H_y(-\alpha))\right] < \theta/2 + \theta/2 = \theta .$$

Finally, the law of iterated projections implies that

$$\eta\left(Y_i^\delta(0)|H_{Y^\delta}(-\alpha)\right) = \eta\left[\eta(y_i(0)|H_y(-\alpha)) \;\middle|\; H_{Y^\delta}(-\alpha)\right] ,$$

so that, given θ, there exists δ' such that for any $\delta < \delta'$

$$d\left[\eta(Y_i^\delta(0)|H_{Y^\delta}(-\alpha)),\ \eta(y_i(0)|H_y(-\alpha))\right] \leq$$

$$\leq d\left[b_\delta , \eta(y_i(0)|H_y(-\alpha))\right] < \theta$$

and we have shown the proposition. ▮

This proposition assures us that the convergence of the α-step-ahead projections obtains under very general conditions. Next we turn to the discussion of the convergence of the MAR.

The first problem we have to deal with when comparing different δ-MAR's is how to normalize them. There are two standard ways of normalizing a discrete MAR: the first requires that $\epsilon(t)$ equals the one-step-ahead innovation, the other is to multiply this innovation by one of the 'square roots' of its variance-covariance matrix, in order to get a white noise in the MAR that has a covariance matrix equal to the identity matrix.

Clearly, if we insist that ϵ^δ be equal to the one-step-ahead innovation (i.e. that $\epsilon^\delta(t) = Y^\delta(t) - \eta(Y^\delta(t)|H_{Y^\delta}(t\delta - \delta)))$ we will not get convergence to the continuous MAR, since $A_0^\delta = I$ for all δ (where I is the identity matrix), and $a(0)$ need not be close to I. Later on we will discuss another important problem that this normalization presents.

On the other hand, we should not insist on normalizing by setting $E(\epsilon^\delta(t) \cdot \epsilon^\delta(t)') = I$ either. In this case, it can be shown that as $\delta \downarrow 0$ the δ-MAR goes to zero for any process.

We propose the following normalization. Let ξ^δ be the one-step-ahead innovation of the δ-process, i.e., $\xi^\delta(t) = Y^\delta(t) - \eta(Y^\delta(t)|H_{Y^\delta}(t\delta - \delta))$. We define the white noise ϵ^δ in (17) as

$$(18) \quad \epsilon^\delta(t) = (\delta)^{\frac{1}{2}} W^{-1} \xi^\delta(t) \quad \text{where } WW' = E\left[(\xi^\delta(t) \cdot \xi^\delta(t)')\right]^{-1} .$$

There are many matrices W that satisfy the above condition. By placing some additional requirements on W, we can resolve this uniqueness problem.[13]

This is a natural normalization in the sense that $\epsilon^\delta(t)$ mimics the properties of a random measure in continuous time (see section 1). Recall that the crucial property of a random measure ζ is that for any interval $\Delta \subset \mathbf{R}$ of length δ, $E(\zeta(\Delta) \cdot \zeta(\Delta)') = \delta \cdot I$. We could interpret ϵ^δ as a random measure defined only on intervals of the form $[k\delta, (k+1)\delta)$, and assigning measure $\epsilon^\delta(k)$ to each of these intervals. Then, the analogue to the above property of ζ is to require that $E(\epsilon^\delta(k) \cdot \epsilon^\delta(k)') = \delta \cdot I$, which is attained with the normalization given by (18).

This is relevant for econometric practice. If one were to compare the MAR's of two series collected at different intervals, the normalization that should be used for each series is the one given by (18).

Define the *function* $A^\delta : \mathbf{R} \to \mathbf{R}^{n \times n}$,

$$\begin{aligned} A^\delta(u) &= A^\delta_k \quad \text{for } u \geq 0 , \quad k \text{ such that } u \in [k\delta, (k+1)\delta) \\ &= 0 \quad \text{for } u < 0 \end{aligned}$$

where A^δ_k corresponds to a given normalization that agrees with (18). In words, A^δ is a step function, and the value that A^δ takes at each step is the value of the MAR coefficient that corresponds to that date.

We will discuss one sense in which A^δ approximates a as $\delta \downarrow 0$. It can be easily checked that:

$$(19) \qquad \int_0^\infty ||A^\delta_i||^2 = \int_0^\infty ||a_i||^2 .$$

This tells us that each row of A^δ belongs to L^2_n, and it has the same norm as the corresponding row of a.

The next proposition states one sense in which A^δ approximates a.

<u>Proposition 7:</u> For any $\alpha > \beta \geq 0$, and any $q \geq 0$:

$$(20) \quad \int_\beta^\alpha A^\delta(u+q) \cdot A^\delta(u)' \, du \to \int_\beta^\alpha a(u+q) \cdot a(u)' \, du \quad \text{as } \delta \downarrow 0 .$$

Proof: Fix α and q; from Proposition 6 and the property of Hilbert spaces mentioned above, the following is true:

$$\text{(21)}\qquad \begin{aligned} \rho_q^\delta &\equiv E\Big[y(q)\cdot[Y^\delta(0)-\eta(Y^\delta(0)|H_{Y^\delta}(-\alpha))]'\Big] \\ &\to E\Big[y(q)\cdot[y(0)-\eta(y(0)|H_y(-\alpha))]'\Big] \equiv \rho \quad \text{as } \delta\downarrow 0\,. \end{aligned}$$

Using the MAR of y (in continuous time) we can write ρ in (21) as

(22)

$$\rho = E\Big[\int_0^\infty a(u)\zeta(q-du)\cdot\Big[\int_0^\alpha a(u)\zeta(0-du)\Big]'\Big] = \int_0^\alpha a(u+q)a(u)'\,du\,.$$

It is enough to show that (20) holds for any sequence $\{\delta_\nu\}$ such that $\delta_\nu\downarrow 0$ as $\nu\to\infty$. Let q_ν be the closest real number to q that is divisible by δ_ν. Let us define ρ^ν by

$$\rho^\nu \equiv E\Big[y(q_\nu)\cdot[Y^{\delta_\nu}(0)-\eta(Y^{\delta_\nu}(0)|H_{Y^{\delta_\nu}}(-\alpha))]'\Big]$$

Since $q_\nu\to q$, $y(q_\nu)\to y(q)$ as $\nu\to\infty$, and it is true that $|\rho_q^{\delta_\nu}-\rho^\nu|\to 0$ as $\nu\to\infty$. Therefore, using (21), $\rho^\nu\to\rho$.

Now, using the δ_ν-discrete MAR of Y^{δ_ν} we can show that

$$\rho^\nu = \sum_{k=0}^{\alpha/\delta_\nu} A_k^{\delta_\nu}\cdot\Big[A_k^{\delta_\nu}+(q_\nu/\delta_\nu)\Big]'\cdot\delta_\nu = \int_0^\alpha A^\delta(u+q)\cdot A^\delta(u)\,du\,.$$

This, (22), and the fact that $\rho^\nu\to\rho$, finishes the proof for the case $\beta=0$.

The case $\beta>0$ is easily shown by noting that for any function f,

$$\int_\beta^\alpha f = \int_0^\alpha f - \int_0^\beta f\,. \qquad \blacksquare$$

The interpretation of this proposition is that the impulse-response function in continuous time of innovations that happen during a fixed period in real time can be approximated by the impulse-response function of innovations over the same period of real time derived from the δ-sampled model.

The results reported in this section so far seem to indicate that with data collected at fine enough intervals, the discrete time model will be a good approximation to the continuous time model. Nonetheless, we have to interpret these results with caution. Propositions 6 and 7 depend crucially on the fact that we fix a period in continuous time first, and then let δ vary. Also, the result in Proposition 7 depends on the normalization given by (18). In the rest of this section, we intend to show how important these points are.

Assume that instead of normalizing ϵ^δ by (18), we set it equal to the one-step-ahead innovation of the δ-sampled process, so that the MAR of Y^δ is now:

$$Y^\delta(t) = \sum_{k=0}^{\infty} A_k \xi^\delta(t-k) \quad \text{for} \quad \xi^\delta(t) = Y^\delta(t) - \eta(Y^\delta(t) | H_{Y^\delta}(t-1)) .$$

It makes no sense to ask whether $\xi^\delta(t)$ converges to some random variable, since $\text{var}(\xi^\delta(t)) \to 0$ as $\delta \downarrow 0$, but we may be interested in determining if the correlation coefficient between ξ^δ and the δ-step-ahead innovation in continuous time ρ_i^δ

$$\rho_i^\delta \equiv \frac{E\left[\xi_i^\delta(t) \cdot \int_0^\delta a_i(u)\zeta(t-du)\right]}{\left[\text{ var }(\xi_i^\delta(t)) \cdot \text{ var }\left(\int_0^\delta a_i(u)\zeta(t-du)\right)\right]^{\frac{1}{2}}}$$

goes to one as $\delta \downarrow 0$. If this convergence does not obtain, it will cast some doubt on the usual practice of interpreting the one-step-ahead innovation as an approximation to the innovation in continuous time over a period of length δ.

We have not yet explored this type of convergence under general conditions, but the next proposition shows that for a certain class of one-dimensional processes in continuous time, ρ^δ does not go to one. We denote the first and second derivatives from the right by $a'(u+)$ and $a''(u+)$.

Proposition 8: Let a be a one-dimensional, continuous time MAR that satisfies:

i) $a(0) = 0$
ii) $a(u) > 0$ for $u > 0$ and u close to zero
iii) $a''(\cdot+)$ exists at $u = 0$
iv) $a'(\cdot-)$ is bounded near zero.

Then ρ^δ does not go to one as $\delta \downarrow 0$.

Proof: (in appendix available from the author).

The set of MAR's in continuous time that satisfy the conditions of this proposition is an important class of models. Imagine that we have a continuous time process with a function a in the MAR that has a first derivative from the right at zero. Then, by unit-averaging this process for all $t \in \mathbf{R}$, we obtain another process that satisfies conditions i) to iv).

7. Conclusion

In what follows, we make an attempt to summarize our results in a non-technical way.

The coefficients of the fundamental moving average representation (MAR) of the discrete, sampled process, are given by

$$A_k = \Big[\int_0^\infty a(u+k)\, c(u)' du\Big] \quad \Big[\int_0^\infty c(u)c(u)' du\Big]^{-1}$$

where a is the MAR in continuous time, and c operates as a weighting function of the continuous MAR. The above equation tells us that when the function c is small on the interval $[1, \infty)$ A_k is mostly an average of a on the interval $[k, k+1)$, and we can consider A_k as a good approximation to a in this sense.

The function c has been thoroughly analyzed in Section 2. There, we have seen that c is large in the interval $[1, \infty)$ when the projection with continuous data is much more accurate than the projection with discrete data.

This restricts largely the type of distortions that can be attributed to temporal aggregation: if an econometrician estimates a MAR that does not agree with what he expects for the particular problem that he is studying, he can ask himself whether the coefficients he observes may have been generated by the above formula, given a function a that *is* of the right form. Also, he may have an *a priori* idea of how good predictions with discrete data are, compared to predictions with continuous data; from this, he can form an *a priori* idea about the function c, and guess how distorted the discrete MAR may be. For example, we were able to show that a systematic effect of time aggregation is to increase the absolute size of the first few coefficients of the MAR.

In Section 5, we have analyzed the problem of unit-averaging data. The formula for the MAR in continuous time that corresponds to unit-averaged data is given by (15′), and it basically implies that this MAR will be smoother, and its mass is shifted one unit to the right, in relation to the MAR and the original process.

Another result of that section is that by mixing variables that are unit-averaged and variables that are sampled in the same discrete time series, we will systematically overstate the importance of the variables that are sampled in determining all the variables in the system. This result is obtained by using the projection approach of section 2, and it illustrates how this approach can be used to derive conclusions that are relevant in econometric practice.

Finally, we study some issues related to the approximation of the model in continuous time by collecting data at finer and finer intervals. We show convergence of the α-step ahead forecast, and we propose a normalization that is convenient for the study of these approximations. Convergence of the discrete time Wold decompositions to the continuous time is studied in Marcet (1987).

Appendix A: Autoregressive Representation of Y

We begin this section by noting that the function h in (7), can not only be approximated by a function of the form $f(u) = \sum \mu_k \, a(u-k)$, for $k\mu_k \in \mathbf{R}^n$, but that it actually is such a function. This is established in the following lemma.

Lemma 1: Given that $i = 1, \ldots, n$, there exist $\lambda_k \in \mathbf{R}^n$ $k = 1, 2, \ldots$ such that:

$$h_i(u) = \sum_{k=1}^{\infty} \lambda'_k \, a(u-k) \quad \text{for almost every} \quad u \in \mathbf{R}$$

where the λ_k's depend on i.

Proof:

By Proposition 1, $h \in A$, so that there exist functions $f_\nu \in L^2_n$, $\nu = 1, 2, \ldots$, such that $f_\nu(u) = \sum_{k=1}^{s_\nu} \mu'_{k,\nu} \, a(u-k)$ for some $\mu_{k,\nu} \in \mathbf{R}^n$, and s_ν finite, and such that $f_\nu \rightarrow h_i$ in the metric of L^2_n as $\nu \rightarrow \infty$. In particular, this implies that

$$\int_1^2 ||f_\nu - f_m||^2 \rightarrow 0 \text{ as } \nu, m \rightarrow \infty \, . \tag{12}$$

Since $a(s) = 0$ for $s < 0$, each f_ν has the property:

$$f_\nu(u) = \mu'_{1,\nu} \, a(u-1) \quad \text{for} \quad u \in [1, 2) \, ,$$

which together with (12) implies that

$$\begin{aligned} &\int_1^2 ||\mu'_{1,\nu} \cdot a(u-1) - \mu'_{1,m} \cdot a(u-1)||^2 = \\ &\int_0^1 ||(\mu'_{1,\nu} - \mu'_{1,m}) \cdot a(u)||^2 \rightarrow 0 \text{ as } \nu, m \rightarrow \infty \, . \end{aligned} \tag{13}$$

Next, we will argue that $\{\mu_1, \nu\}$ converges in $\mathbf{R}^n$ as $\nu \to \infty$. We can rewrite (13) as:

$$(\mu_{1,\nu} - \mu_{1,m})' \cdot \left[\int_0^1 a(u)a(u)'du\right] \cdot (\mu_{1,\nu} - \mu_{1,m}) \to 0 .$$

We claim that for any positive definite matrix C, and any sequence $\{x_n\}$ in $\mathbf{R}^n$, $x_n C x_n \to 0$ if and only if $x_n \to 0$ in $\mathbf{R}^n$.[14] But the assumption that y is a full rank process in continuous time implies that $\int_0^1 aa'$ is positive definite, because this is the covariance matrix of the one-step-ahead innovation of y in continuous time. Therefore, (13) implies that $\mu_{1,\nu} \to \lambda_1$ as $\nu \to \infty$ for some $\lambda_1 \in \mathbf{R}^n$.

It is easy to show that, if $\mu_{k,\nu} \to \lambda_k$ for all $k = 1, \ldots, r$ for a finite integer r, then $\mu_{r+1,\nu} \to \lambda_{r+1}$ as $\nu \to \infty$. To see this, note that

$$\int_0^{r+1} ||f_\nu - f_m||^2 \to 0 \text{ as } \nu, m \to \infty ;$$

by the inductive assumption, this implies that $\int_0^1 ||(\mu_{r+1,\nu} - \mu_{r+1,m}) \cdot a(u)||^2 \to 0$, and $\mu_{r+1,\nu} \to \lambda_{r+1} \in \mathbf{R}^n$ as $\nu \to \infty$.

Now, choose any integer K. Clearly

$$\int_0^K ||f_\nu(u) - \sum_{k=1}^{\infty} \lambda_k a(u-k)||^2 du \to 0 \text{ and } \int_0^K ||f_\nu - h_i||^2 \to 0 \text{ as } \nu \to \infty .$$

Since the limits are unique, $h_i(u) = \sum_{k=1}^{\infty} \lambda_k' a(u-k)$ for almost every $u < K$. Therefore, this equality holds for almost every $u \in \mathbf{R}$. ∎

If we apply this lemma to equation (7), we can write Y as

$$Y_i(t) = \varepsilon(t) + \int_1^{\infty} \sum_{k=1}^{\infty} \lambda_k' a(u-k)\zeta(t-du)$$

so that if we can pull the summation outside the integral sign, we can write:

$$\text{(14)} \qquad Y_i(t) = \varepsilon_i(t) + \sum_{k=1}^{\infty} \lambda_k' Y(t-k)$$

and in this case Y has an autoregressive representation, with the coefficients of this representation being the λ's in the above lemma.

To "pull the summation sign outside the integral" in fact means that the following is true:

$$\int_1^{\infty} \sum_{k=1}^{\infty} \lambda_k' a(u-k)\zeta(t-du) = \lim_{m\to\infty} \sum_{k=1}^{m} \lambda_k' \int_1^{\infty} a(u-k)\zeta(t-du)$$

which, by definition of convergence in mean square, means:

$$\lim_{m\to\infty} \int_1^\infty ||\sum_{k=1}^{\infty} \lambda'_k a(u-k) - \sum_{k=1}^{m} \lambda'_k a(u-k)||^2 du = 0, \text{ or}$$

$$\lim_{m\to\infty} \int_1^\infty ||\sum_{k=m}^{\infty} \lambda'_k a(u-k)||^2 du = 0 \ .$$

The next lemma gives a sufficient condition for the interchange of summation and integration.

Lemma 2: If the sequence $\{\lambda_k\}$ is absolutely summable (element by element), then (14) holds.

Proof:

By applying twice the triangle inequality and pulling the constants λ out of the norm in L_n^2, we can justify the following steps:

$$\left[\int_1^\infty ||\sum_{k=m}^{\infty} \lambda'_k a(u-k)||^2\right]^{-\frac{1}{2}} \leq \sum_{j=1}^{n} \left[\int_1^\infty ||\sum_{k=m}^{\infty} \lambda^j_k a_j(u-k)||^2\right]^{-\frac{1}{2}} \leq$$

$$\sum_{j=1}^{n} \sum_{k=m}^{\infty} |\lambda^j_k| \left[\int_1^\infty ||a_j(u-k)||^2\right]^{-\frac{1}{2}} \leq K \sum_{j=1}^{n} \sum_{k=m}^{\infty} |\lambda^j_k| \ ,$$

where K is constant. If λ_k is absolutely summable the right hand side goes to zero, and by our earlier comment, (14) will hold. ▮

Notes

1. Rozanov (1967) contains the definitions of linear regularity, stationarity, random measures, integration with respect to random measures, and a proof of the existence of the MAR in continuous and discrete time. Throughout the paper, we will only use orthonormal random measures, so that for any pair Δ, δ' of disjoint Borel subsets of R, $E(\zeta_i(\Delta))^2 = |\Delta|$ for all i, $E(\zeta_j(\Delta) \cdot \zeta_i(\Delta)) = 0$ if $i \neq j$, and for any i, j, $E(\zeta_j(\Delta) \cdot \zeta_j(\Delta')) = 0$.

2. See Rozanov (1967) for a definition of spectral density and its properties. Equation (5) is known as the folding formula. In most of our discussion, it is quite clear how our results could be generalized to the case of a non-full rank process y, but allowing for this case

would complicate our notation without adding much substance to the problem.

3. See Rozanov (1967), page 3.

4. Rozanov (1967) observes that the spaces $H_\varepsilon(\infty)$ (for ε an uncorrelated random measure) and L_n^2 are unitarily isometric, because:

$$\| \int f d\varepsilon - \int g d\varepsilon \| = \|f - g\|$$

where the first norm is for random variables, and the second represents the norm in L_n^2. This tells us that $H_\varepsilon(\infty)$ and L_n^2 "are arranged in the same way" (page 3, Rozanov (1967)); therefore, in view of equations (1) and (7), we would expect that there was a relationship between h and a in terms of the topology of L_n^2. This relationship is precisely what Proposition 1 uncovers.

5. See Rudin, theorems 12.3 and 12.4.

6. Here, we have used: $\varepsilon(t) = \int_0^\infty c\, d\zeta = Y(t) - \eta(Y - i(t) | H_Y(t-1)) = \int_0^\infty (a - h) d\zeta$. We also have used the fact that for any $f, g \in L_n^2$, if $\int f d\varepsilon = \int g d\varepsilon$, then var $[\int f d\varepsilon - \int g d\varepsilon] = 0$, so that $\int \| f - g \|^2 = 0$ and $f = g$ almost everywhere. Therefore, $c = a - h$.

7. Since $f(u) = \sum_{k=1}^\infty \sum_{j=1}^n \lambda_{k,j}\, a_j(u-k)$, if $u < 1$ then $u - k < 0$ for any of the k's in the summation; $a_j(u-k) = 0$ and $f(u) = 0$.

8. Sims (1972b) shows that Y_2 does not Granger cause Y_1 iff $\eta(Y_2(t) \mid H_{Y_1}(t)) = \eta(Y_2(t) \mid H_{Y_1}(\infty))$. But from Sims (1971), it is clear that, in general, this equality will not hold for the sampled process, even if it holds for the underlying continuous time process.

9. See Wheeden and Zygmund (1977) for the definition of indefinite integral and its properties.

10. See Wheeden and Zygmund, (1977) page 101.

11. To consider a correct example, take $a_{11}(u) = e^{-\lambda u}$, $a_{21}(u) = Ke^{-\lambda u}$ for all $u \geq 0$, and $a_{12} \equiv 0$. In continuous time, the second variable does not Granger cause the first one and, by Proposition 4, the same is true in the sampled process. However it can be shown that if we unit-average the first variable, the second variable will Granger cause the first one.

12. We have used the package RATS for finding these decompositions. In all the simulations, the covariance matrix of the innovations is $\begin{bmatrix} 1. & .2 \\ .2 & 1. \end{bmatrix}$. After aggregating, the length of each discrete series was 135 observations. We have run many other simulations; the results reported in the tables are fairly representative of all the simulations we have run. The result in table 2 is the "worst" we could attain in all the simulations we ran.

13. One option is to require that W be the Cholesky decomposition of the covariance matrix of ξ^{δ}. Another is to make $W = H \cdot \Lambda^{\frac{1}{2}}$, where H has the eigenvectors of this covariance matrix in its columns, and $\Lambda^{\frac{1}{2}}$ has the square root of the eigenvalues on its diagonal, zeros elsewhere.

14. If C is a positive definite, $n \times n$ matrix, there exists a positive definite matrix W such that $W'W = C$, so that $(Wx_n)'(Wx_n) \to 0$, which is equivalent with $Wx_n \to 0$ in $\mathbf{R}^n$. Since W is invertible, it is true that $x_n \to 0$ in $\mathbf{R}^n$.

References

Anderson, B.D.O. (1978), "Second-Order Convergent Algorithms for the Steady-State Riccati Equation," *International Journal Control*, 28, 295–306.

Anderson, B.D.O. and J.B. Moore (1979), *Optimal Filtering*, Prentice Hall, Inc., N.J.

Ansley, C.F., and R. Kohn (1983), "Exact Likelihood of Vector Autoregressive-Moving Average Process with Missing or Aggregated Data," *Biometrika*, 70, 275–278.

Barro, R. J. (1977), "Unanticipated Money Growth and Unemployment in the United States," *American Economic Review*, 67.

Barro, R.J. (1979), "On the Determination of the Public Debt," *Journal of Political Economy*, 87, 940–71.

Becker, G.S. and K.M. Murphy (1988), "A Theory of Rational Addiction," *Journal of Political Economy*, 96, 675–700.

Beltrami, E.J. and M.R. Wohlers (1966), *Distributions and the Boundary Values of Analytic Functions*, Academic Press, New York.

Bernanke, B. (1985), "Adjustment Costs, Durables, and Aggregate Consumption," *Journal of Monetary Economics*, 15, 41–68.

Bernanke, B. (1986), "Alternative Explorations of the Money-Income Correlation," in *Real Business cycles, Real Exchange Rates, and Actual Policies*, ed. Karl Brunner and Allan H. Meltzer. Carnegie-Rochester Conference Series on Public Policy, 25, 49–99. Amsterdam: North-Holland.

Bergstrom, A.R. (1976), *Statistical Inference in Continuous Time Economic Models*, Amsterdam: North-Holland.

Bergstrom, A.R. (1983), "Gaussian Estimation of Structural Parameters in Higher Order Continuous Time Dynamic Models," *Econometrica*, 51, 117–152.

Blinder, A.S. and A. Deaton (1985), "The Time Series Consumption Function Revisited," *Brookings Paper on Economic Activity*, 2, 465–511.

Breeden, D.T. (1979), "An Intertemporal Asset Pricing Model with Stochastic Consumption and Investment Opportunities," *Journal of Financial Economics*, 7, 265–296.

Brock, W.A. and M.J.P. MaGill (1979), "Dynamics Under Uncertainty," *Econometrica* 47, 843–868.

Campbell, J.Y. (1986), "Bond and Stock Returns in a Simple Exchange Model," *Quarterly Journal of Economics*, 101, 785–803.

Campbell, J.Y. (1987), "Does Saving Anticipate Declining Labor Income? An Alternative Test of the Permanent Income Hypothesis," *Econometrica*, 55, 1249–1273.

Campbell, J.Y. and R.J. Shiller (1987), "Cointegration and Tests of Present Value Models," *Journal of Political Economy*, 95, 1062–1088.

Campbell, J.Y. and R.J. Shiller (1988), "Dividend-Price Ratios and Expectations of Future Dividends and Discount Factors," *Review of Financial Studies*, 1, 195–228.

Christiano, L.J. (1980), "Notes on Factoring Continuous Time Rational Spectral Densities," manuscript, Federal Reserve Bank of Minneapolis.

Christiano, L.J. (1984), "The Effects of Aggregation over Time on Tests of the Representative Agent Model of Consumption," manuscript.

Christiano, L.J. and M.S. Eichenbaum (1985), "A Continuous Time, General Equilibrium, Inventory-Sales Model," manuscript.

Christiano, L. J. and M.S. Eichenbaum (1986), "Temporal Aggregation and Structural Inference in Macroeconomics," Carnegie-Rochester Conference on Public Policy, Vol. 26.

Christiano, L.J., M.S. Eichenbaum, and D. Marshall (1990), "The Permanent Income Hypothesis Revisited," manuscript. Forthcoming in *Econometrica*.

Churchill, R.V., J.W. Brown and R.F. Verhey (1974), *Complex Variables and Applications*, 3rd edition, New York: McGraw Hill.

Coddington, E. A. and N. Levinson (1955), *Theory of Ordinary Differential Equations*, New York: McGraw-Hill.

Constantinides, G.M. (1990), "Habit Formation: A Resolution of the Equity Premium Puzzle," *Journal of Political Economy*, 98, 519–543.

Cox, J.D., J.E. Ingersol, Jr., and S.A. Ross (1985), "A Theory of the Term Structure of Interest Rates," *Econometrica*, 53, 129–151.

DeJong, D.N., J.C. Nankervis, N.E. Savin, and C.H. Whiteman (1988), "Integration versus Trend-Stationarity in Macroeconomic Time Series," University of Iowa Department of Economics Working Paper #88-27, December.

DeJong, D.N. and C.H. Whiteman (1989a), "The Temporal Stability of Dividends and Stock Prices: Evidence from the Likelihood Function," University of Iowa Department of Economics Working Paper #89-03, February.

DeJong, D.N. and C.H. Whiteman (1989b), "Trends and Random Walks in Macroeconomic Time Series: A Reconsideration Based on the Likelihood Principle," University of Iowa Department of Economics Working Paper #89-04, February.

Denman, E.D. and A.N. Beavers (1976), "The Matrix Sign Function and Computations in Systems," *Applied Mathematics and Computation*, 2, 63–94.

Dunsmuir, W. (1978), "A Central Limit Theorem for Parameter Estimation in Stationary Vector Time Series and its Application to Models for Signal Observed with Noise," *Annals of Statistics*, 7, 490–506.

Dunsmuir, W., and E.J. Hannan (1976), "Vector Linear Time Series Models," *Advances in Applied Probability*, 8, 339–364.

Eckstein, Z. (1984), "Rational Expectations Modeling of Agricultural Supply," *Journal of Political Economy*, 93, 1–19.

Eichenbaum, M.S. (1983), "A Rational Expectations Equilibrium Model of Finished Goods and Employment," *Journal of Monetary Economics*, 12, 259–278.

Eichenbaum, M.S. and L.P. Hansen (1990), "Estimating Models with Intertemporal Substitution Using Aggregate Time Series Data," *Journal of Business and Economic Statistics*, 8, 53–69.

Emre, E. and O. Huseyin (1975), "Generalization of Leverrier's Algorithm to Polynomial Matrices of Arbitrary Degree," *IEEE Transactions on Automatic Control*, February.

Engle, R.F., and C.W.J. Granger (1987), "Cointegration and Error-Correction: Representation, Estimation and Tesging," *Econometrica*, 55, 251–276.

Engle, R. and M. Watson (1981), "A One Factor Multivariate Time Series Model of Metropolitan Wage Rates," *Journal of the American Statistical Association*, 76, 774–781.

Flavin, M.A. (1981), "The Adjustment of Consumption to Changing Expectations About Future Income," *Journal of Political Economy*, 89, 974–1009.

Futia, C.A. (1981), "Rational Expectations in Stationary Linear Models," *Econometrica*, 49, 171–192.

Gantmacher, F.R. (1959), *The Theory of Matrices*, Vol. 1, New York: Chelsea.

Geweke, J.B. (1977a), "Wage and Price Dynamics in U.S. Manufacturing," in *New Methods in Business Cycle Research: Proceedings From a Conference*, ed. C.A. Sims, 111–56. Minneapolis: Federal Reserve Bank of Minneapolis.

Geweke, J.B. (1977b), "The Dynamic Factor Analysis of Economic Time Series," in D.J. Aigner and A.S. Goldberger, eds., *Latent Variables in Socio-Economic Models*, Ch. 19, 365–383. Amsterdam: North-Holland.

Geweke, J.B. (1978), "Temporal Aggregation in the Multivariate Regression Model," *Econometrica*, 46, 643–662.

Geweke, J.B. and K.J. Singleton (1981a), "Maximum Likelihood 'Confirmations' Factor Analysis of Economic Time Series," *International Economic Review*, 22, 37–54.

Geweke, J.B. and K.J. Singleton (1981b), "Latent Variable Models for Time Series: A Frequency Domain Approach with an Application to the Permanent Income Hypothesis," *Journal of Economics*, 17, 287–304.

Gohberg, I., P. Lancaster and L. Rodman (1982), *Matrix Polynomials*, Academic Press.

Gould, J.P. (1968), "Adjustment Costs in the Theory of Investment of the Firm," *Review of Economic Studies*, 35, 47–56.

Granger, C.W.J. (1969), "Investigating Causal Relations by Econometric Models and Cross-Spectral Methods," *Econometrica*, 37, 424–438.

Griliches, Z. (1967), "Distributed Lags: A Survey," *Econometrica*, 35, 16–40.

Grossman, S., A. Melino, and R.J. Shiller (1987), "Estimating the Continuous Time Consumption-Based Asset Pricing Model," *Journal of Business and Economic Statistics*, 6, 315–327.

Hakkio, C.S. and M. Rush (1986), "Co-integration and the Government's Budget Deficit," Working Paper, Federal Reserve Bank of Kansas City.

Hall, R.E. (1978), "Stochastic Implications of the Life Cycle-Permanent Income Hypothesis: Theory and Evidence," *Journal of Political Economy*, 86, 971–987.

Hall, R.E. (1988), "Intertemporal Substitution and Consumption," *Journal of Political Economy*, 96, 339–357.

Halmos, P.R., (1950), *Measure Theory*, New York, Van Nostrand.

Halmos, P.R., (1957), *Introduction to Hilbert Space*, New York, Chelsea Publishing Company.

Hamilton, J.D. and M.A. Flavin (1986), "On the Limitations of Government Borrowing: a Framework for Empirical Testing," *American Economic Review* 76, 808–819.

Hannan, E. J. (1970), *Multiple Time Series*, Wiley, New York.

Hansen, L.P. (1980), "Large Sample Properties of Generalized Method of Moments Estimators," Carnegie-Mellon University, Manuscript.

Hansen, L.P. (1987) "Calculating Asset Prices in Three Example Economics," in T.F. Bewley, ed., *Advances in Econometrics*, Vol. 4, Cambridge University Press, 207–243.

Hansen, L.P., D. Epple and W. Roberds (1985), "Linear Quadratic Models of Resource Depletion," in T.J. Sargent, ed., *Energy, Foresight, and Strategy*. Resources for the Future, Washington, D.C., 101–142.

Hansen, L.P. and R.J. Hodrick (1980), "Forward Exchange Rates as Optimal Predictors of Future Spot Rates: An Econometric Analysis," *Journal of Political Economy*, 88, 829–853.

Hansen, L.P. and R.J. Hodrick (1983), *Exchange Rates and International Macroeconomics* NBER, J. Frenkel, ed., 113–152.

Hansen, L.P. and T.J. Sargent (1980a), "Formulating and Estimating Dynamic Linear Rational Expectations Models," *Journal of Economic Dynamics and Control*, 2, 7–46.

Hansen, L.P. and T.J. Sargent (1980b), "Methods for Estimating Continuous Time Rational Expectations Models from Discrete Time Data," Staff Report 59, Federal Reserve Bank of Minneapolis.

Hansen, L.P. and T.J. Sargent (1981a), "Linear Rational Expectations Models for Dynamically Interrelated Variables," R.E. Lucas, Jr., and T.J. Sargent eds. *Rational Expectations and Econometric Practice*, Minneapolis: University of Minnesota Press.

Hansen, L.P. and T.J. Sargent (1981b), "A Note of Wiener-Kolmogorov Prediction Formulas for Rational Expectations Models," *Economic Letters*, 8, 255–260.

Hansen, L.P. and T.J. Sargent (1981c), "Identification of Continuous Time Rational Expectations Models from Discrete Time Data," manuscript.

Hansen, L.P. and T.J. Sargent (1981d) "Formulating and Estimating Continuous Time Rational Expectations Models," Staff Report 75, Federal Reserve Bank of Minneapolis, October.

Hansen, L.P. and T.J. Sargent (1981e), "Exact Linear Rational Expectations Models: Specification and Estimation," Research Department Staff Report 71, Federal Reserve Bank of Minneapolis.

Hansen, L.P. and T.J. Sargent (1982), "Instrumental Variables Procedures for Estimating Linear Rational Expectations Models," *Journal of Monetary Economics*, 9, 263–296.

Hansen, L.P. and T.J. Sargent (1983a), "Aggregation over Time and the Inverse Optimal Predictor Problem for Adaptive Expectations in Continuous Time," *International Economic Review*, 24, 1–20.

Hansen, L.P. and T.J. Sargent (1983b), "The Dimensionality of the Aliasing Problem in Models with Rational Spectral Densities," *Econometrica*, 51, 377–387.

Hansen, L. P. and T.J. Sargent (1984) "Two Difficulties in Interpreting Vector Autoregressions," Working Paper 227, Federal Reserve Bank of Minneapolis.

Hansen, L.P. and T.J. Sargent (1990), "Recursive Linear Models of Dynamic Economies," manuscript.

Hansen, L.P. and K.J. Singleton (1982), "Generalized Instrumental Variables Estimators of Nonlinear Rational Expectations Models," *Econometrica*, 50, 1269–1286.

Hansen, L.P. and K.J. Singleton (1983), "Stochastic Consumption, Risk Aversion, and the Temporal Behavior of Asset Returns," *Journal of Political Economy*, 91, 249–265.

Harvey, A.C. and J.H. Stock (1985), "The Estimation of Higher Order Continuous Time Autoregressive Models," *Journal of Econometric Theory*, 1, 97–112.

Harvey, A.C. and J.H. Stock (1987), "Estimating Integrated Higher Order Continuous Time Autoregressions with an Application to Money-income Causality," Working Paper No. E 87-28. Stanford, CA: Hoover Institution Press.

Harvey, C. (1988), "The Real Term Structure of Consumption Growth," *Journal of Financial Economics*, 22, 305–333.

Hatanaka, M. (1975), "On the Global Identification of the Dynamic Simultaneous Equations Model with Stationary Disturbances," *International Economic Review*, 16, 545–554.

Hayashi, F. and C.A. Sims (1983), "Nearly Efficient Estimation of Time Series Models with Predetermined, but not Exogenous Instruments," *Econometrica*, 51, 3, 783–798.

Heaton, J. (1989), "The Interaction Between Time-Nonseparable Preferences and Time Aggregation," University of Chicago Ph.D. Dissertation.

Jones, R.H. (1980), "Maximum Likelihood Fitting of ARMA Models to Time Series with Missing Observations," *Technometrics*, 22, 389–

395.

Kennan, J. (1979), "The Estimation of Partial Adjustment Models with Rational Expectations," *Econometrica*, 47, 1441–1456.

Kennan, John (1988), "An Econometric Analysis of Fluctuations in Aggregate Labor Supply and Demand," *Econometrica*, 56, 2, 317–333.

Kollintzas, T. (1985), "The Symmetric Linear Rational Expectations Model," *Econometrica*, 53, 963–985.

Kwakernaak, H. and R. Sivan (1972), *Linear Optimal Control Systems*, New York, Wiley.

LeRoy, S.F. (1973), "Risk Aversion and the Martingale Property of Stock Prices," *International Economic Review*, 14, 436–446.

LeRoy, S.F. and R. Porter (1981), "The Present Value Relation: Tests Based on Implied Variance Bounds," *Econometrica*, 49, 97–113.

Litterman, R. and L. Weiss (1985), "Money, Real Interest Rates and Output: A Reinterpretation of Post-war Data," *Econometrica*, 53, 129–156.

Lo, A. (1988), "Maximum Likelihood Estimation of Generalized Ito Processes with Discretely Sampled Data," *Journal of Econometric Theory*, 4, 231–247.

Lucas, R.E., Jr. (1967), "Adjustment Costs and the Theory of Supply," *Journal of Political Economy*, 75, 321–334.

Lucas, R.E., Jr. (1972), "Econometric Testing of the Natural Rate Hypothesis," in *The Econometrics of Price Determination Conference*, ed. by O. Eckstein. Washington: Board of Governors of the Federal Reserve System.

Lucas, R.E., Jr. (1973), "Some International Evidence on Output-Inflation Tradeoffs," *American Economic Review* 63 (June): 326–34.

Lucas, R.E., Jr. (1976), "Econometric Policy Evaluation: A Critique," in K. Brunner and A.H. Meltzer, eds., *The Phillips Curve and Labor Markets*, Carnegie-Rochester Conference on Public Policy 1: 19–46, Amsterdam: North Holland.

Lucas, R.E., Jr. (1978), "Asset prices in an Exchange Economy," *Econometrica* 46, 1429–1445.

Lucas, R.E., Jr. (1981), "Optimal Investments with Rational Expectations," *Rational Expectations and Econometric Practice*, Vol. 1, 1981, 127–156.

Lucas, R.E., Jr. and E.C. Prescott (1971) "Investment Under Uncertainty," *Econometrica*, 39, 659–681.

Lucas, R.E., Jr. and T.J. Sargent (1981), *Introductory Essay to Rational Expectations and Econometric Practice*, Minneapolis: University of Minnesota Press.

Luenberger, D.G., *Optimization by Vector Space Methods*, John Wiley & Sons, 1969.

Mankiw, N.G. (1982), "Hall's Consumption Hypothesis and Durable Goods," *Journal of Monetary Economics*, 10, 417–425.

Marcet, A. (1987), "Approximation of the Continuous Wold Decomposition by Frequent Sampling," mimeo, Carnegie Mellon University.

Miller, P.J. and W. Roberds (1987), "The Quantitative Significance of the Lucas Critique," Federal Reserve Bank of Minneapolis Staff Report #109.

Monahan, J. (1984), "A Note on Enforcing Stationarity in Autoregressive-Moving Average Models," *Biometrika*, 71, 403–404.

Mortenson, D.T. (1973), "Generalized Costs of Adjustment and Dynamic Factor Demand Theory," *Econometrica*, 41, 657–665.

Muth, J. (1960), "Optimal Properties of Exponentially Weighted Forecasts," *Journal of the American Statistical Association*, 55, 299–306.

Muth, J. (1961), "Rational Expectations and the Theory of Price Movements," *Econometrica*, 29, 315–335.

Nerlove, M. (1967), "Distributed Lags and Unobserved Components in Economic Time Series," in *Ten Economic Studies in the Tradition of Irving Fisher*, ed. by W. Fellner, et. al., New York: Wiley.

Phillips, A.W. (1959), "The Estimation of Parameters in Systems of Stochastic Differential Equations," *Biometrika*, 59, 67–76.

Phillips, P.C.B. (1972), "The Structural Estimation of a Stochastic Differential Equation System," *Econometrica*, 40, 1021–1041.

Phillips, P.C.B. (1973), "The Problem of Identification in Finite Parameter Continuous Time Models," *Journal of Econometrics*, 1, 351–362.

Phillips, P.C.B. (1974), "The Estimation of Some Continuous Time Models," *Econometrica*, 42, 803–823.

Pollak, R.A. (1970), "Habit Formation and Dynamic Demand Functions," *Journal of Political Economy*, 78, 745–763.

Roberts, J.D. (1971), "Linear Model Reduction and Solution of Algebraic Riccati Equations by use of the Sign Function," CUED/B-Control/TR-13 Report, Cambridge University.

Robinson, P.M. (1977), "The Construction and Estimation of Continuous Time Models and Discrete Approximations in Econometrics,"

Journal of Econometrics, 6, 173–197.

Royden, H.L. (1968), *Real Analysis*, New York, MacMillan Publishing Co., Inc.

Rozanov, Y.A. (1967), *Stationary Random Processes*, Holden-Day: San Francisco.

Rubinstein, M. (1976), "The Valuation of Uncertain Income Streams and the Pricing of Options," *Bell Journal of Economics*, 7, 407–425.

Rudin, W. (1974), *Real and Complex Analysis*, McGraw Hill: New York.

Ryder, H.E. and G.M. Heal (1973), "Optimal Growth with Intertemporally Dependent Preferences," *Review of Economic Studies*, 40, 1–31.

Sargent, T.J. (1971), "A Note on the 'Accelerationist' Controversy," *Journal of Money, Credit, and Banking*, 3, 721–725.

Sargent, T.J. (1978), "Estimation of Dynamic Labor Demand Schedules Under Rational Expectations," *Journal of Political Economy*, 86, 1009–1044.

Sargent, T.J. (1979), "A Note on Maximum Likelihood Estimation of the Rational Expectations of the Term Structure," *Journal of Monetary Economics*, 5, 133–143.

Sargent, T.J. (1981), "Interpreting Economic Time Series," *Journal of Political Economy*, 89, 213–248.

Sargent, T.J. (1983), "Preliminary Introduction to Continuous Time Stochastic Processes," manuscript.

Sargent, T.J. (1987a), *Dynamic Macroeconomic Theory*, Cambridge: Harvard University Press.

Sargent, T.J. (1987b), *Macroeconomic Theory*, 2nd ed., New York: Academic Press.

Sargent, T.J. and C.A. Sims (1977), "Business Cycle Modeling Without Pretending to Have too Much *A Priori* Economic Theory." In *New Methods for Business Cycle Research* (C.A. Sims, ed.), Minneapolis: Federal Reserve Bank.

Schwarz, Z.G. (1978), "Estimating the Dimension of a Model," *Annals of Statistics*, 6, 461–464.

Shiller, R.J. (1972), "Rational Expectations and the Structure of Interest Rates," Ph.D. Dissertation (Massachusetts Institute of Technology, Cambridge, MA).

Shiller, R.J. (1979), "The Volatility of Long-Term Interest Rates and Expectations Models of the Term Structure," *Journal of Political*

Economy, 87, 1190–1219.

Shiller, R.J. (1981), "Do Stock Prices Move too much to be Justified by Subsequent Changes in Dividends?" *The American Economic Review*, 71, 421–436.

Shim, S.D. (1984), "Inflation and the Government Budget Constraint: International Evidence," Ph.D. Thesis, University of Minnesota.

Sims, C.A. (1971), "Discrete Approximations to Continuous Time Lag Distributions in Econometrics," *Econometrica*, 39, 545–564.

Sims, C.A. (1972a), "Approximate Prior Restriction on Distributed Lag Estimation," *Journal of the American Statistical Association*, 67, 169–175.

Sims, C.A. (1972b), "Money, Income and Causality," *American Economic Review*, 62, 540–552.

Sims, C.A. (1974), "Distributed Lags," in *Frontiers of Quantitative Economics*, Vol. II, ed. by M.D. Intrilligator and D.A. Kendrick, Amsterdam: North-Holland, 289–332.

Sims, C.A. (1980), "Macroeconomics and Reality," *Econometrica*, 62, 540–552.

Sims, C.A. (1984), "Martingale-Like Behavior of Prices and Interest Rates," Department of Economics, University of Minnesota, working paper, October.

Sims, C.A. (1988), "Bayesian Skepticism on Unit Root Econometrics," *Journal of Economic Dynamics and Control*, 12, 463–474.

Sims, C.A. and H. Uhlig (1990), "Understanding Unit Rooters: A Helicopter Tour," forthcoming in *Econometrica*.

Stigler, G.J. and G.S. Becker (1977), "De Gustibus Non Est Disputandum," *American Economic Review*, 67, 76–90.

Sundaresan, M. (1989), "Intertemporally Dependent Preferences and the Volatility of Consumption and Wealth," *The Review of Financial Studies*, 2, 73–89.

Telser, L.G. and R.J. Graves (1972), *Functional Analysis in Mathematical Economics*, Chicago: University of Chicago Press.

Townsend, R.M. (1983), "Forecasting the Forecasts for Others," *Journal of Political Economy*, 91, 546–588.

Treadway, A.B. (1969), "On Rational Entrepreneur Behavior and the Demand for Investment," *Review of Economic Studies*, 367, 227–240.

Trehan, B. and C.E. Walsh (1988), "Common Trends, the Government Budget Constraints, and Revenue Smoothing," *Journal of Economic Dynamics and Control*, 12, 425–446.

Vaughan, D.R. (1969), "A Negative-Exponential Solution to the Matrix Riccati Equation," *IEEE Transactions on Automatic Control*, 14, 72–75.

Watson, M. and R. Engle (1983), "Alternative Algorithms for the Estimation of Dynamic Factor, MIMIC and Varying Coefficient Regression Models," *Journal of Econometrics*, 23, 385–400.

West, K.D. (1988), "The Insensitivity of Consumption to News About Income," *Journal of Monetary Economics*, 21, 17–33.

Wheeden, R.L. and A. Zygmund (1977), *Measure and Integral; An Introduction to Real Analysis*, Marcel Dekker Inc.: New York.

Whiteman, C. (1983), *Linear Rational Expectations Models: A User's Guide*, University of Minnesota Press.

Wilcox, D.W. (1989), "What Do We Know about Consumption?" Federal Reserve Board manuscript.

Contributors

Lars Peter Hansen, University of Chicago

John Heaton, Massachusetts Institute of Technology

Albert Marcet, Carnegie-Mellon University and Universitat Pompeu Fabra

William Roberds, Federal Reserve Bank of Atlanta

Thomas J. Sargent, Hoover Institution, Stanford University